The Experimental Side of Modeling

Minnesota Studies in the Philosophy of Science

ALSO IN THIS SERIES

THE EXPERIMENTAL SIDE
OF MODELING

ISABELLE F. PESCHARD
AND BAS C. VAN FRAASSEN, EDITORS

Minnesota Studies in the Philosophy of Science 21

 University of Minnesota Press

Minneapolis

London

Published by the University of Minnesota Press
111 Third Avenue South, Suite 290
Minneapolis, MN 55401-2520
http://www.upress.umn.edu

Printed in the United States of America on acid-free paper

The University of Minnesota is an equal-opportunity educator and employer.

Library of Congress Cataloging-in-Publication Data
Names: Peschard, Isabelle F., editor. | Van Fraassen, Bas C., editor.
Title: The experimental side of modeling / edited by Isabelle F. Peschard
    and Bas C. van Fraassen.
Description: Minneapolis : University of Minnesota Press, [2018] | Series: Minnesota
    studies in the philosophy of science ; 21 | Includes bibliographical references and index. |
Identifiers: LCCN 2018001583 (print) | ISBN 978-1-5179-0533-0 (hc) |
    ISBN 978-1-5179-0534-7 (pb)
Subjects: LCSH: Science—Experiments—Methodology. | Science—Methodology. |
    Experimental design.
Classification: LCC Q182.3 .E96285 2018 (print) | DDC 001.4/34—dc23
LC record available at https://lccn.loc.gov/2018001583

UMP BmB 2018

CONTENTS

PREFACE AND ACKNOWLEDGMENTS

In a major shift during the twentieth century, philosophy of science changed its focus from scientific theories to models, measurement, and experimentation. But certain traditional concerns were slow to give way to the new directions of research thus opened. Models are a form of scientific representation, and experimental activity serves in many cases to test empirical implications against data. The focus placed almost entirely on these aspects alone, however, was soon seen as narrow and by itself inadequate in relation to scientific practice.

Three workshops under the title *The Experimental Side of Modeling* were held at San Francisco State University in 2009–11. In these sessions, extant philosophical approaches to modeling and experiment met intense debate and severe critique. The conceptions of model as representation and of experiment as tribunal were challenged in explorations of precisely how modeling is related to experimentation, how these two activities are intertwined in practice, how disparities between experimental results and the results of simulations of models are adjudicated, and what evaluational criteria apply, beyond assessment in terms of accuracy of representation. The main issues addressed included the construction of models in conjunction with experimentation in a scientific inquiry, with specific case studies in various scientific fields; the status of measurement and the function of experiment in the identification of relevant parameters; the consequent reconception of the phenomenon and of what is to be accounted for by a model; and the interplay between experimenting, modeling, and simulation when the results do not mesh.

THANKS ARE DUE to all the other participants in the workshops. In addition to the contributors to this volume, the presenters included Karen Barad, Mieke Boon, Elizabeth Lloyd, Alan Love, Roberta Millstein, Seppo Poutanen, and Alison Wylie. The main commentators were David Stump, Ásta Kristjana Sveinsdottir, Martin Thomson-Jones, and Rasmus Winther. Participating as commentator-at-large were Arthur Fine, James Griesemer, Edward Mac-Kinnon, Elizabeth Potter, Shannon Vallor, and Andrea Woody.

Professor Anita Silvers, chair of the San Francisco State University Philosophy Department, provided the center of energy for support and organization of the workshops; Dean Paul Sherwin provided the financial and material resources that made it all possible; and Brandon Hopkins of Nousoul Digital Publishers helped with the graphics. To these organizers and participants, to the university itself, and to the many graduate students who willingly lent a hand with logistics and technical support, we owe our heartfelt gratitude.

INTRODUCTION

ISABELLE F. PESCHARD AND BAS C. VAN FRAASSEN

The philosophical essays on modeling and experimenting in this volume were originally presented at three workshops on the experimental side of modeling at San Francisco State University in 2009–2011. The debates that began there continue today, and our hope is that they will make a difference to what the philosophy of science will be. As a guide to this collection the introduction will have two parts: an overview of the individual contributions, followed by an account of the historical and methodological context of these studies.

OVERVIEW OF THE CONTRIBUTIONS

In this overview we present the individual essays of the collection with a focus on the relations between them and on how they can be read within a general framework for current studies in this area. Thus, first of all, we do not view experiments simply as a tribunal for producing the data against which models are tested, but rather as themselves designed and shaped in interaction with the modeling process. Both the mediating role of models and the model dependence, as well as theory dependence, of measurement come to the fore. And, conversely, the sequence of changing experimental setups in a scientific inquiry will, in a number of case studies, be seen to make a constructive contribution to the process of modeling and simulating. What will be especially evident and significant is the new conception of the role of data not as a given, but as what needs to be identified as what is to be accounted for. This involves normative and conceptual innovation: the phenomena to be investigated and modeled are not precisely identified

1

beforehand but specified in the course of such an interaction between experimental and modeling activity.

It is natural to begin this collection with Ronald Giere's "Models of Experiments" because it presents the most general, overall schema for the description of scientific modeling and experimenting: a model of modeling, one might say. At the same time, precisely because of its clarity and scope, it leads into a number of controversies taken on by the other participants.

Ron Giere's construal of the process of constructing and evaluating models centers on the comparison between models of the data and fully specified representational models. Fully specified representational models are depicted as resulting from other models. They are obtained from more general representational models by giving values to all the variables that appear in those models. This part of Giere's *model of the modeling process* (briefly, his *meta-model*) represents the components of a theoretical activity, but it also comprises the insertion of empirical information, the information required to give values to the variables that characterize a specific experimental situation.

Giere's account gives a rich representation of the different elements that are involved in the activities of producing, on the one hand, theoretical models and, on the other hand, data models. It also makes clear that these activities are neither purely theoretical nor purely empirical.

In Giere's meta-model, data models are obtained from models of experiments. It should be noted, however, that they are not obtained from models of experiments in the same sense as fully specified representational models are obtained from representational models. Strictly speaking, one should say that data models are obtained from a transformation of what Giere calls "recorded data," data resulting from measuring activity, which itself results from running an experiment. And these operations also appear in his meta-model. But a model of the experiment is a representation of the experimental setup that produces the data from which the data model is constructed. So data models are obtained from models of experiments in the sense that the model of the experiment presents a source of constraint on what kind of data can be obtained, and thereby a constraint on the data model. It is this constraint that, in Giere's model of the modeling process, the arrow from the model of the experiment to the data model represents.

Another way to read the relation represented on Giere's schema between data model and model of the experiment is to see it as indicating that understanding a data model requires understanding the process that produced it, and thereby understanding the model of the experiment.

In the same way as the activity involved in producing theoretical models for comparison with data models is not exclusively theoretical, because it requires the use of empirical information, the activity involved in producing data models is not exclusively empirical. According to Giere, a model of the experiment is a map of the experimental setup that has to include three categories of components: *material* (representation of the instruments), *computational* (computational network involved in producing models of the data from instruments' input), and *agential* (operations performed by human agents). Running the experiment, from the use of instruments to the transformation of data resulting in the data model, requires a certain understanding of the functioning of the instruments and of the computational analysis of the data which will be, to a variable extent, theoretical.

Because of the generality and large scope of Giere's meta-model, it leaves open many topics that are taken up later on in the collection. One might find it surprising that the notion of phenomenon does not appear here and wonder where it could be placed. One possibility would be to locate it within "the world." But in Giere's meta-model, the world appears as something that is given, neither produced nor constrained by any other component of the modeling process. As we will see in Joseph Rouse's essay, to merely locate phenomena within the world "as given" betrays the normative dimension that is involved in recognizing something as a phenomenon, which is very different from recognizing something as simply part of the world. What one might find missing also in Giere's model is the dynamical, interactive relation between the activities that produce theoretical and data models, which is emphasized in Anthony Chemero's contribution, and again in Michael Weisberg's. In Giere's schema, these activities are only related through their product when they are compared. Joe Rouse's contribution calls that into question as well in his discussion of the role of phenomena in conceptual understanding.

Thus, we see how Giere's presentation opens the door to a wide range of issues and controversies. In Giere's schema, the model of the experiment, just like the world, and like phenomena if they are mere components of the world, appears as given. We do not mean, of course, that Giere is oblivious to how the model of the experiment comes about; it is not created ex nihilo. But in this schema it is left aside as yet how it is produced or constrained by other components of the modeling process. It is given, and it is descriptive: a description of the experimental setup. But scientists may be wrong in their judgment that something is a phenomenon. And they may be wrong, or on the contrary show remarkable insight, in their conception of what needs and what does not need to be measured to further our understanding of a certain object of investigation. Arguably, just like for phenomena, and to allow for the possibility of error, the conception of the model of the experiment needs to integrate a normative dimension. In that view, when one offers a model of the experiment, one is not simply offering a description but also making a judgment about what needs to be, what should be, measured. As we will see, the addition of this normative dimension seems to be what most clearly separates Chemero's from Giere's approach to the model of the experiment.

In Anthony Chemero's "Dynamics, Data, and Noise in the Cognitive Sciences," a model of the experiment is of a more general type than it is for Giere. In Chemero's sense it is not a specific experimental arrangement but rather a kind of experimental arrangement, defined in terms of what quantities it is able to measure, without specifying the details of what instruments to use or how.

Chemero comes to issues in the philosophy of modeling and its experimental components from the perspective of dynamical modeling in cognitive science. This clear and lucid discussion of dynamical modeling combines with an insightful philosophical reflection on some of the Chemero's own experimental work. The essay makes two main contributions of particular relevance to our overall topic, both at odds with traditional views on the experimental activity associated with model testing. The first one is to provide a clear counterexample to what Chemero calls "the methodological truism": "The goal of a good experiment is to maximize primary variance, eliminate secondary variance, and minimize error variance."

The primary variance is the evolution of the variable of interest, or systematic variance of the dependent variable, under the effect of some other, independent variables that are under control. The secondary variance is the systematic variance of an independent variable resulting from the effect of variables other than the independent variables. The error variance is what is typically referred to as "noise": unsystematic variance of the dependent variable. The methodological truism asserts, in effect, that noise is not informative and is only polluting the "real" signal, the real evolution of the variable of interest. So it should be minimized and what remains will be washed out by using some measure of the central tendency of the collection of measurements of the dependent variable, such as the mean.

As counterexample to the methodological truism, Chemero presents a case study where the noise is carrying information about the very phenomenon under investigation.

Chemero's second aim is to illustrate a normative notion of the model of the experiment and, at the same time, to show that to construct a good experimental setup is not just a matter of technical skills but also a matter of determining what quantities to measure to obtain empirical evidence for a target phenomenon. Along the way it is also shown that dynamical models in cognitive science are not simply descriptions of phenomena but fully deserve to be considered as explanations as well.

The case study that Chemero analyzes uses a Haken-Kelso-Bunz model of coordination dynamics. This model has been famously used to model the dynamics of finger wagging, predicting, for instance, that beyond a certain rate only one of two possible behaviors is stable, namely in-phase movement. But it has also been applied to many other topics of study in cognitive science, from neural coordination dynamics to speech production to interpersonal coordination. Chemero discusses the use of this model to investigate the transition described by Heidegger between *readiness-to-hand* and *unreadiness-to-hand*: "The dynamical model was used to put Heidegger's claims about phenomenology into touch with potentially gatherable data."

Given the project to test Heidegger's idea of a transition between two forms of interactions with tools—ready-to-hand and unready-to-hand—the challenge was to determine what sort of quantities could serve as evidence for this transition. To determine what the experimental setup needs to be is to offer a model of the experiment. To offer such a model is not just to give a description of experimental arrangements, it is to make a claim about what sort of quantities can serve as evidence for the phenomenon of interest and thus need to be measured. It is a normative claim; and as Rouse's essay will make clear, it is defeasible. Scientists may come to revise their view about what sorts of quantities characterize a phenomenon—and that is especially so in cognitive science.

The experimenters arrive at a model of the experiment by first interpreting the notion of ready-to-hand in terms of an interaction-dominant system: to be in a ready-to-hand form of interaction with a tool is to form an interaction-dominant system with the tool. This means a kind of system where "the interactions are more powerful than the intrinsic dynamics of the components" and where the components of the system cannot be treated in isolation. An interaction-dominant system exhibits a special variety of fluctuation called *1/f noise* or *pink noise* ("a kind of not-quite-random, correlated noise"), and the experimenters take it that evidence for interaction dominance will be evidence for the ready-to-hand form of interaction.

That is how the methodological truism is contradicted: in that case, noise is not something that is simply altering the informative part of the signal, not something that should be minimized and be compensated for. Instead, it is a part of the signal that is carrying information, information about the same thing that the systematic variation of the primary variable carries information about. As we will see below, Knuuttila and Loettgers make a similar point in their essay in the context of experimental modeling of biological mechanisms.

Interpretation and representation are, it is said, two sides of one coin. St. Elmo's fire may be represented theoretically as a continuous electric discharge; observations of St. Elmo's fire provide data that may be interpreted as signs of a continuous electric discharge. To see St. Elmo's fire as a phenomenon to be modeled as plasma—ionized air—rather than as fire is to see it through the eyes of theory; it is possible only if, in the words of Francis Bacon, "experience has become literate." But the relation between theory and world is not unidirectional; the phenomena, Joseph Rouse argues in his essay, contribute to conceptual understanding.

Rouse's contribution addresses the question of what mediates between model and world by focusing on the issue of conceptual understanding. The link between model and world is clarified as the argument unfolds to show that the phenomena play this mediating role. This may be puzzling at first blush: how can the phenomena play a role in conceptual understanding, in the development and articulation of concepts involved in explaining the world with the help of models? According to Rouse both what conceptual articulation consists in, and the role that phenomena play in this articulation, have been misunderstood.

Traditionally viewed, conceptual articulation is a theoretical activity. About the only role that phenomena play in this activity is as that which this activity aims to describe. Phenomena are seen as what theories are about and what data provide evidence for (Bogen and Woodward 1988). When models and world are put in relation with one another, the concepts are already there to be applied and phenomena are already there to be explained. As Rouse says, "[the relation between theory and the world] comes into philosophical purview only after the world has already been conceptualized." And the aim of experimental activity, from this perspective, is to help determine whether concepts actually apply to the world. Issues related to experimental activity are seen as mainly technical, as a matter of creating the causal conditions prescribed by the concepts and models that we are trying to apply.

Rouse's move away from this view on conceptual articulation as mainly theoretical, on phenomena as always already conceptualized, and on experimental activity as a technical challenge starts with giving to phenomena the role of mediator between models and world. Phenomena are not some more or less hidden parts of the world that will hopefully, with enough luck and technical skill, be discovered. Instead, they contribute to the conceptual development and articulation through which we make the world's complexity intelligible.

How do phenomena play this mediating role? First of all, phenomena could not be mediators if they were merely objects of description (or prediction or explanation): "Phenomena *show* something important about the world, rather than our merely *finding* something there." Phenomena's mediating role starts with and is grounded in the recognition of their significance. The recognition of their significance is not a descriptive, empirical judgment. It is a normative judgment. It is as normative as saying that a person is good, and just as open to revision. The question of whether they are discovered or constructed is pointless. It is like asking whether the meter unit is discovered or constructed. What matters is the normative judgment that makes the one-meter length into a unit. Similarly, with phenomena what matters is the judgment of their significance, that they show something important about the world.

Saying that something is a phenomenon is to make a bet: what we are betting is what is at stake in making this normative judgment. We are betting that it will enable us to see new things that we were not able to see before— new patterns, new regularities. That is the important thing about the world that the phenomenon is taken to show: the possibility of ordering the world in terms similar to it, and to use these terms to create intelligibility, order, and to make some new judgments of similarity and differences, in the same way as we do with the meter unit.

Rouse makes clear that normativity does not imply self-vindication. On the contrary, we may have to recognize that what we took for a phenomenon does not actually show anything important about the world. That is why there is something at stake in judging that something is a phenomenon. There would be nothing at stake if the judgment were not revisable.

So how do phenomena contribute to conceptual articulation? One tempting mistake at this point would be to think that they do so by providing the conditions in which concepts are applied. We cannot reduce a concept to what is happening under some specific conditions. Phenomena may serve as an anchor for concepts, an epicenter, but they point beyond themselves at the larger world, and that is where the domain of application of concepts lies. Once a pattern has been recognized as significant ("outer recognition"), the issue is "how to go on rightly"—that is, how to correctly identify other cases as fitting this same pattern, as belonging to the domain of application of the same concept ("inner recognition").

Rouse refuses to reduce the domain of application of concepts to the conditions in which phenomena are obtained, but these conditions are important to his story. To understand how phenomena can play a role without becoming a trap, a black hole for concepts that would exhaust their content, we need to introduce the notion of an experimental system. Perhaps the most important aspect of experimental systems is that they belong to a historical and dynamic network of "systematically interconnected experimental capacities." What matters, in the story of how phenomena contribute to conceptual articulation, "is not a static experimental setting, but its ongoing differential reproduction, as new, potentially destabilizing elements are introduced into a relatively well-understood system." It is the development of this experimental network that shows how concepts can be applied further and how conceptual domains can be articulated.

Crucially then, in Rouse's account, in the same way as there is no way to tell in advance how the experimental network will extend, there is no way to tell in advance how concepts will apply elsewhere, how they will extend further. But what is certain is that the concepts that we have make us answerable in their terms to what will happen in the course of the exploration of new experimental domain: "Concepts commit us to more than we know how to say or do." In some cases, as experimental domains extend, the difficulties may be regarded as conceptually inconsequential, but in other cases they may require a new understanding of the concepts we had, to "conceptually re-organize a whole region of inquiry."

Tarja Knuuttila and Andrea Loettgers present an exciting and empirically informed discussion of a new form of mediation between models and experiments. Their subject is the use of a new kind of models, synthetic models, which are systems built from genetic material and implemented in a natural cell environment.

The authors offer an in-depth analysis of the research conducted with two synthetic models, the Repressilator and a dual feedback oscillator. In both cases the use of these models led to new insights that could not have been obtained with just mathematical models and traditional experiments on model organisms. The new insights obtained are distinct contributions of intrinsic and extrinsic noise in cell processes in the former case and of an unanticipated robustness of oscillatory regime in the latter.

Elsewhere on the nature of synthetic models, Knuuttila and Loettgers contrast their functioning and use to that of another genetically engineered system, genetically modified *Escherichia coli* bacteria (Knuuttila and Loettgers 2014). As they explain, this system was used as a measuring system by virtue of its noise-sensing ability. By contrast, the synthetic models are synthetic systems capable of producing a behavior.

After a review of the debate on the similarities and dissimilarities between models and experiments, and using their case study in support, the authors argue that the epistemic function that synthetic models are able to play is due to characteristics they share both with models and with experiments. Thus, the discussion by Knuuttila and Loettgers of synthetic systems sheds a new light on the debate concerning the similarities and dissimilarities between models and experiments. They acknowledge some similarities but highlight certain differences, in particular the limited number of the theoretical model's components and the specific materiality of the experimental system, on a par with the system under investigation.

What makes the synthetic systems so peculiar is that some of the characteristics they share with models and with experimental systems are those that make models and experiments epistemically dissimilar. What such systems share with models is that by contrast with natural systems they are constructed on the basis of a model—a mathematical model—and their structure only comprises a limited number of interactive components. What they share with experiments is that when placed in natural conditions they are able to produce unanticipated behavior: they start having "a life of their own." And it is precisely by virtue of this life of their own that they could lead to new insights when their behavior is not as expected.

It may be surprising to think that the construction of a system in accordance with a theoretical model could be seen as something new. After all, experimental systems that are constructed to test a model are constructed according to the model in question. (That is what it is, in Nancy Cartwright's suggestive phrase, to be a nomological machine.) But the object of investigation and source of insight in the experiments described by Knuuttila and Loettgers are not the behavior of the synthetic model per se but rather its behavior in natural conditions. Such a synthetic system is not a material model, constructed to be an object of study and a source of claims about some other system, of which it is supposed to be a model. The point of the synthetic system is to be placed in natural conditions, and the source of surprise and insight is the way it behaves in such conditions. What these systems then make possible is to see how natural conditions affect the predicted behavior of the model. That is possible because, by contrast with the mathematical model used as a basis to produce it, the synthetic model is made of the same sort of material as appears in those natural conditions. Synthetic models are also different from the better known "model organisms." The difference is this: whereas the latter have the complexity and opacity of natural systems, the former have the simplicity of an engineered system. Like models and unlike organisms, they are made up of a limited number of components. And this feature, also, as the authors make clear, is instrumental to their innovative epistemic function.

The philosophical fortunes of the concept of causation in the twentieth century could be the stuff of a gripping novel. Russell's dismissal of the notion of cause in his "Mysticism and Logic" early in the twentieth century was not prophetic of how the century would end. Of the many upheavals in the story, two are especially relevant to the essays here. One was the turn from evidential decision theory to causal decision theory, in which Nancy Cartwright's "Causal Laws and Effective Strategies" (1979) was a seminal contribution. The other was the interventionist conception of causal models in the work of Glymour, Pearl, and Woodward that, arguably, had become the received view by the year 2000. In the essays by Cartwright and by Jenann Ismael, we will see the implications of this story for both fundamental philosophical categories and practical applications.

Nancy Cartwright addresses the question of what would be good evidence that a given policy will be successful in a given situation. Will the results of implementing the policy in a certain situation serve as a basis to anticipate the results of the implementation in the new situation? Her essay discusses and compares the abilities of experiments and models to provide reliable evidence. Note well that she is targeting a specific kind of experiment—randomized control trial (RCT) experiments—and a specific kind of model: "purpose-built single-case causal models."

For the results of the implementation of a policy to count as good evidence for the policy's efficacy, the implementation needs to meet the experimental standards of a well-designed and well-implemented RCT. But even when that is the case, the evidence does not transfer well, according to Cartwright, to a new situation.

A well-designed and well-implemented RCT can, in the best cases, *clinch* causal conclusions about the effect of the treatment on the population that received the treatment, the study population. That is, if the experiment comprises two populations that can be considered identical in all respects except for receiving versus not receiving the treatment, and if a proper measurement shows an effect in the study population, then the effect can be reliably attributed to the causal effect of the treatment. The question is whether these conclusions, or conclusions about the implementation of a policy in a given situation on a given population, can travel to a new target population/situation.

According to Cartwright, what the results of a RCT can show is that *in the specific conditions* of the implementation the implementation makes a difference. But it says nothing about the difference it might make in a new situation. For, supposing the implementation was successful, there must be several factors that are contributing, directly or indirectly, in addition to the policy itself, to the success of that implementation. Cartwright calls these factors *support factors*. And some of these factors may not be present in the new situation or other factors may be present that will hamper the production of the outcome.

One might think that the randomization that governs the selection of the populations in the RCT provides a reliable basis for the generalization of results of the experiment. But all it does, Cartwright says, is ensure that there is no feature that could *by itself* causally account for the difference in results between the populations. That does not preclude that some features shared by the two populations function as support factors in the study population and made the success of the treatment possible. All that success requires is that some members of the population exhibit these features in such a way as to produce a higher average effect in the study population. Given that we do not know who these members are, we are unable to identify the support factors that made the treatment successful in these individuals.

Cartwright contrasts RCT experiments with what she calls purpose-built single-case causal models. A purpose-built single-case causal model will, at minimum, show what features/factors are necessary for the success of the implementation: the support factors. The most informative, though, will be a model that shows the contribution of the different factors necessary for the success of the implementation of the policy in a given situation/population, which may not be present in the new situation.

It might seem that the purpose-built single-case causal model is, in fact, very general because it draws on a general knowledge we have about different factors. The example that Cartwright specifically discusses is the implementation of a policy aiming to improve learning outcomes by decreasing the size of classes. The implementation was a failure in California whereas it had been successful in Tennessee. The reduction of class size produces a need for a greater number of classrooms and teachers; in California, that meant hiring less qualified teachers and using rooms for classes that had previously been used for learning activities. A purpose-built single-case causal model, says Cartwright, could have been used to study, say, the relationship between quality of teaching and quality of learning outcome. The model would draw on general knowledge about the relationship between quality of teaching and quality of learning outcome, but it would be a single case in that it would only represent the support factors that may or may not be present in the target situation. That maintaining teaching quality while increasing sufficiently the number of teachers would be problematic in California is the sort of fact, she says, that we hope the modeler will dig out to produce the model we must rely on for predicting the outcomes of our proposed policy.

But how, one might wonder, do we evaluate how bad it really is to lower the number of learning activities? How do we determine that the quality of the teaching will decrease to such an extent that it will counteract the benefit of class reduction? How do we measure teaching quality in the first place?

According to Cartwright, we do not "have to know the correct functional form of the relation between teacher quality, classroom quality, class size, and educational outcomes to suggest correctly . . . that outcomes depend on a function that . . . increases as teacher and room quality increase and decreases as class size increases." Still, it seems important, and impossible in the abstract, to evaluate the extent to which the positive effect expected from class reduction will be hampered by the negative effect expected from a decrease in teachers or room quality or number of learning activities.

Cartwright recognizes that an experiment investigating the effect of the policy on a representative sample of the policy target would be able to tell us what we want to know. But it is difficult to produce, in general, a sample that is representative of a population. And, furthermore, as the California example makes clear, the scale at which the policy is implemented may have a significant effect on the result of the implementation.

Without testing the implementation of the policy itself, can we not think of other, less ambitious experiments that would still be informative? Would it be helpful, for instance, to have some quantitative indication of the difference that learning activities or lack of experience of teachers, alone, makes on learning outcomes? If we can hope to have scientists good enough to produce models that take into account just the right factors, could we not hope to have scientists good enough to conceive of experiments to investigate these factors so as to get clearer on their contribution?

Cartwright takes it that "responsible policy making requires us to estimate as best we can what is in this model, within the bounds of our capabilities and constraints." But aren't experiments generally the way in which precisely we try to estimate the contribution of the factors that we expect to have an effect on the evolution of the dependent variable that we are interested in?

Cartwright's discussion gives good reasons to think that the results of an experiment that is not guided by a modeling process might be of little use beyond showing what is happening in the specific conditions in which it is realized. It is not clear, however, that models without experiment can give us the quantitative information that is required for its being a guide for action and decision.

She notes that "the experiment can provide significant suggestions about what policies to consider, but it is the model that tells us whether the policy will produce the outcome for us." Experiments may not be able to say, just by themselves, what the support factors are nor whether they are present in the target situation/population. But, it seems they may help a good deal once we have an idea of what the support factors are, to get clearer on how these factors contribute to the production of the effect. And they may even be of help in finding some factors that had not been considered but that do, after all, have an effect.

Jenann Ismael insists that philosophers' focus on the products of science rather than its practice has distorted specifically the discussion of laws and causality. She begins with a reflection on Bertrand Russell's skeptical 1914 paper on the notion of cause, which predisposed a whole generation of empiricists against that notion, but Ismael sees the rebirth of the concept of causal modeling as starting with Nancy Cartwright's 1979 argument that causal knowledge is indispensable in the design of effective strategies.

Russell had argued that the notion of cause plays no role in modern natural science, that it should be abandoned in philosophy and replaced by the concept of a global dynamical law. His guiding example was Newton's physics, which provided dynamical laws expressed in the form of differential equations that could be used to compute the state of the world at one time as a function of its state at another. Russell's message that the fundamental nomic generalizations of physics are global laws of temporal evolution is characterized by Ismael as a prevailing mistaken conviction in the philosophy of science during much of the twentieth century.

But, as she recounts, the concept of cause appears to be perfectly respectable in the philosophy of science today. While one might point to the overlap between the new focus on modality in metaphysics, such as in the work of David Lewis, and the attention to this by such philosophers of science as Jeremy Butterfield and Christopher Hitchcock, Ismael sees the turning point in Nancy Cartwright's placing causal factors at the heart of practical decision making. Cartwright's argument pertained most immediately to the controversy between causal and evidential decision theory at that time. But it also marks the beginning of the development of a formal framework for representing causal relations in science, completed in the "interventionist" account of causal modeling due to Clark Glymour, Judea Pearl (whom Ismael takes as her main example), and James Woodward.

Does causal information outrun the information generally contained in global dynamical laws? To argue that it does, Ismael focuses on how a complex system is represented as a modular collection of stable and autonomous components, the "mechanisms." The behavior of each of these is represented as a function, and interventions are local modifications of these functions. The dynamical law for the whole can be recovered by assembling these in a configuration that imposes constraints on their relative variation so as to display how interventions on the input to one mechanism propagate throughout the system. But the same evolution at the global level could be realized through alternative ways of assembling mechanisms, hence it is not in general possible to recover the causal information from the global dynamics.

Causality and mechanism are modal notions, as much as the concept of global laws. The causal realism Ismael argues for involves an insistence on the priority of mechanism over law: global laws retain their importance but are the emergent product of causal mechanisms at work in nature.

Eventually, after all the scientific strife and controversy, we expect to see strong evidence brought to us by science, to shape practical decisions as well as worldviews. But just what is evidence, how is it secured against error, how is it weighed, and how deeply is it indebted for its status to what is accepted as theory? In the process that involves data-generating procedures, theoretical reasoning, model construction, and experimentation, finally leading to claims of evidence, what are the normative constraints? The role of norms and values in experimentation and modeling, in assessment and validation, has recently taken its place among the liveliest and most controversial topics in our field.

Deborah Mayo enters the fray concerning evidential reasoning in the sciences with a study of the methodology salient in the recent "discovery" of the Higgs boson. Her theme and topic are well expressed in one of her subheadings: "Detaching the Inferences from the Evidence." For as we emphasized earlier, the data are not a given; what counts as evidence and how it is to be understood are what is at issue.

When the results of the Higgs boson detection were made public, the findings were presented in the terminology of orthodox ("frequentist") statistics, and there were immediate critical reactions by Bayesian statisticians. Mayo investigates this dispute over scientific methodology, continuing her long-standing defense of the former (she prefers the term "error statistical") method by a careful inquiry into how the Higgs boson results were (or should be) formulated. Mayo sees "the main function of statistical method as controlling the relative frequency of erroneous inferences in the long run." There are different types of error, which are all possible ways of being mistaken in the interpretation of the data and the understanding of the phenomenon on the basis of the data, such as mistakes about what factor is responsible for the effect that is observed. According to Mayo, to understand how to probe the different components of the experiment for errors—which includes the use of instruments, their manipulation and reasoning, and how to control for errors or make up for them—is part of what it is to understand the phenomenon under investigation.

Central to Mayo's approach is the notion of severity: a test of a hypothesis H is severe if not only does it produce results that agree with H if H is correct, but it also very likely produces results that do not agree with H if H is not correct. Severe testing is the basis of what Mayo calls "argument from

error," the argument that justifies the experimenter's trust in his or her interpretation of the data. To argue from error is to argue that a misinterpreta-
tion of the data would have been revealed by the experimental procedure.

In the experiments pertaining to the Higgs boson, a statistical model is
presented of the detector, within which researchers define a "global signal
strength" parameter μ, such that $\mu = 0$ corresponds to the detection of the
background only (hypothesis H_0), and $\mu = 1$ corresponds to the Standard
Model Higgs boson signal in addition to the background (hypothesis H).
The statistical test records differences in the positive direction, in standard
deviation or sigma units. The improbability of an excess as large as 5 sigma
alludes to the *sampling distribution* associated with such signal-like results
or "bumps," fortified with much cross-checking. In particular, the probability of observing a result as extreme as 5 sigma, under the assumption it was
generated by background alone—that is, assuming that H_0 is correct—is approximately 1 in 3,500,000.

Can this be summarized as "the probability that the results were just a
statistical fluke is 1 in 3,500,000"? It might be objected that this fallaciously
applies the probability to H_0 itself—a posterior probability of H_0. Mayo argues that this is not so.

The conceptual distinctions to be observed are indeed subtle. H_0 does
not say the observed results are due to background alone, although *if H_0 were
true* (about what is generating the data), it follows that various results would
occur with specified probabilities. The probability is assigned to the observation of such large or larger bumps (at both sites) on the supposition that
they are due to background alone. These computations are based on simulating what it would be like under H_0 (*given a detector model*). Now the inference actually *detached* from the evidence is something like *There is strong
evidence for H*. This inference does indeed rely on an implicit principle of
evidence—in fact, on a variant of the *severe* or stringent testing requirement
for evidence. There are cases, regrettably, that do commit the fallacy of
"transposing the conditional" from a low significance level to a low posterior to the null. But in the proper methodology, what is going on is precisely
as in the case of the Higgs boson detection. The difference is as subtle as it is
important, and it is crucial to the understanding of experimental practice.

Eric Winsberg's "Values and Evidence in Model-Based Climate Forecast-
ing" is similarly concerned with the relation between evidence and inference, focusing on the role of values in science through a cogent discussion of
the problem of uncertainty quantification (UQ) in climate science—that is,

the challenging task of attaching a degree of uncertainty to climate models predictions. Here the role of normative decisions becomes abundantly clear.

Winsberg starts from a specific argument for the ineliminable role of social or ethical values in the epistemic endeavors and achievements of science, the inductive risk argument. It goes like this: given that no scientific claim is ever certain, to accept or reject it is to take a risk and to make the value judgment that it is, socially or ethically, worth taking. No stretch of imagination is needed to show how this type of judgment is involved in controversies over whether, or what, actions should be taken on the basis of climate change predictions.

Richard Jeffrey's answer to this argument was that whether action should be taken on the basis of scientific claims is not a scientific issue. Science can be value-free so long as scientists limit themselves to estimating the evidential probability of their claims and attach estimates of uncertainties to all scientific claims to knowledge. Taking the estimation of uncertainties of climate model predictions as an example, Winsberg shows how challenging this probabilistic program can be, not just technically but also, and even more importantly, conceptually. In climate science, the method to estimate uncertainties uses an ensemble of climate models obtained on the basis of different assumptions, approximations, or parametrizations. Scientists need to make some methodological choices—of certain modeling techniques, for example—and those choices are not, Winsberg argues, always independent of non-strictly-epistemic value judgments. Not only do the ensemble methods rely on presuppositions about relations between models that are clearly incorrect, especially the presupposition that they are all independent or equally reliable, but it is difficult to see how it could be possible to correct them. The probabilistic program cannot then, Winsberg concludes, produce a value-free science because value judgments are involved in many ways in the very work that is required to arrive at an estimation of uncertainty. The ways they are involved make it impossible to trace these value judgments or even to draw a clear line between normative and epistemic judgments. What gets in the way of drawing this line, in the case of climate models, is that not only are they complex objects, with different parts that are complex and complexly coupled, but they also have a complex history.

Climate models are a result of a series of methodological choices influencing "what options will be available for solving problems that arise at a later time" and, depending on which problems will arise, what will be the model's epistemic successes or failures. What Joseph Rouse argues about concepts applies here to methodological choices: we do not know what they commit us to. This actual and historical complexity makes it impossible to determine beforehand the effects of the assumptions and choices involved in the construction of the models. For this very reason, the force of his claims, Winsberg says, will not be supported by a single, specific example of value judgment that would have influenced the construction or assessment of climate models. His argument is not about some value judgments happening here or there; it is about how pervasive such judgments are, and about their "entrenchment"—their being so intimately part of the modeling process that they become "mostly hidden from view . . . They are buried in the historical past under the complexity, epistemic distributiveness, and generative entrenchment of climate models."

While Winsberg's essay focuses on the interplay of values and criteria in the evaluation of models in climate science, Michael Weisberg proposes a general theory of model validation with a case study in ecology.

What does it mean that a model is good or that one model is better than another one? Traditional theories of confirmation are concerned with the confirmation of claims that are supposed to be true or false. But that is exactly why they are not appropriate, Michael Weisberg argues, for the evaluation of idealized models—which are not, and are not intended to be, truthful representations of their target. In their stead, Weisberg proposes a theory of model validation.

Weisberg illustrates this problem of model evaluation with two models in contemporary ecology. They are models of a beech forest in central Europe that is composed of regions with trees of various species, sizes, and ages. The resulting patchy pattern of the forest is what the models are supposed to account for. Both models are cellular automata: an array of cells that represents a patch of forest of a certain kind and a set of transition rules that, step by step, take each of the cells from one state to another. The transition rules determine at each step what each cell represents based on what the cell and the neighboring cells represented at the previous step. One of the models is more sophisticated than the other: the representation and transition rules take into account a larger number of factors that characterize the forest (e.g., the height of the trees). It is considered a better model than the other one, but it is still an idealization; and the other one, in spite of its being simpler, is nevertheless regarded as at least partially validated and also explanatory.

Weisberg's project is to develop an account of model validation that explains the basis on which a model is evaluated and what makes one model better than another one. His account builds on Ronald Giere's view that it is similarities between models and real systems that make it possible for scientists to use models to represent real systems. Following Giere, Weisberg proposes to understand the validation of models as confirmation of theoretical hypotheses about the similarity between models and their intended targets. Weisberg's account also follows Giere in making the modeler's purpose a component of the evaluation of models: different purposes may require different models of the same real system and different forms of similarities between the model and the target.

In contrast to Giere, who did not propose an objective measure of similarity, Weisberg's account does. The account posits a set of features of the model and the target, and Weisberg defines the similarity of a model to its target as "a function of the features they share, penalized by the features they do not share"; more precisely, "it is expressed as a ratio of shared features to non-shared features." Weisberg's account is then able to offer a more precise understanding of "scientific purpose." For example, if the purpose is to build a causal model then the model will need to share the features that characterize the mechanism producing the phenomenon of interest. If the purpose includes simplicity, the modeler should minimize in the model the causal features that are not in the target. By contrast, if the purpose is to build a "how-possible" model, what matters most is that the model shares the static and dynamic features of the target. The way in which the purpose influences the evaluation appears in the formula through a weighting function that assigns a weight to the features that are shared and to those that are not shared. The weight of these features expresses how much it matters to the modeler that these features are shared or not shared and determines the way in which their being shared or not shared influences the evaluation.

This account enables Weisberg to explain why it may not be possible to satisfy different purposes with a single model and why some trade-offs might be necessary. It also enables him to account for what makes, given a purpose, one model better than another one.

And it enables him to account for another really important and neglected aspect of modeling and model evaluation: its iterative aspect. As we saw, the formula includes a weighting function that ascribes weights to the different features to express their relative importance. The purpose of the modeler indirectly plays a role in determining the weighting function in that it determines what the model needs to do. But what directly determines, given the purpose, what the weighting function has to be is some knowledge about what the model has to be like in order to do what the purpose dictates. In some cases, this knowledge will be theoretical: to do this or that the model should have such and such features and should not or does not need to have such and such other features. In some cases, however, Weisberg points out, there will be no theoretical basis to determine which features are more important. Instead, the weights may have to be iteratively adjusted on the basis of a comparison between what the model does and the empirical data that need to be accounted for, until the model does what it is intended to do. That will involve a back and forth between the development of the model and the comparison with the data.

Weisberg speaks here of an interaction between the development of the model and the collection of data. Such an interaction suggests that the development of the model has an effect on the collection of the data. It is an effect that is also discussed in other contributions (Chemero, Knuuttila and Loettgers, and Rouse). It is not clear, at first, how that would happen under Weisberg's account because it seems to take the intended target, with its features, and the purpose for granted in determining the appropriate validation formula. But it does not need to do so. The model may suggest new ways to investigate the target, for example, by looking for some aspects of the phenomenon not yet noticed but predicted by the model. Finally, Weisberg discusses the basis on which a validated model can be deemed a reliable instrument to provide new results.

In addition to the validation of the model, what is needed, Weisberg explains, is a robustness analysis that shows that the results of the model are stable through certain perturbations. The modelers generally not only want the model to provide new results about the target it was tested against but also want these results to generalize to similar targets. Weisberg does not seem to distinguish between these expectations here. In contrast, Cartwright's essay makes clear that to project the results of a model for one situation to a new situation may require much more than a robustness analysis.

Symposium on Measurement

Deborah Mayo rightly refers to measurement operations as data-generating procedures. But a number or text or graphics generating procedure is not necessarily generating data: what counts as data, as relevant data, or as evidence, depends on what is to be accounted for in the relevant experimental inquiry or to be represented in the relevant model. If a specific procedure is a measurement operation then its outcomes are values of a specific quantity, but whether there is a quantity that a specific procedure measures, or which quantity that is, and whether it is what is represented as that quantity in a given model, is generally itself what is at stake.

In this symposium, centering on Paul Teller's "Measurement Accuracy Realism," the very concept of measurement and its traditional understanding are subjected to scrutiny and debate.

Van Fraassen's introduction displays the historical background in philosophy of science for the changing debates about measurement, beginning with Helmholtz in the nineteenth century. The theory that held center stage in the twentieth was the representational theory of measurement developed by Patrick Suppes and his coworkers. Its main difficulties pertained to the identification of the quantities measured. The rival analytic theory of measurement proposed by Domotor and Batitsky evades this problem by postulating the reality of physical quantities of concern to the empirical sciences. Both theories are characterized, according to van Fraassen, by the sophistication of their mathematical development and paucity of empirical basis. It is exactly the problem of identifying the quantities measured that is the target of Paul Teller's main critique, which goes considerably beyond the traditional difficulties that had been posed for those two theories of measurement.

First of all, Teller challenges "traditional measurement-accuracy realism," according to which there are in nature quantities of which concrete systems have definite values. This is a direct challenge to the analytic theory of measurement, which simply postulates that. But the identification of quantities through their pertinent measurement procedures, on which the representationalist theory relied, is subject to a devastating critique of the disparity between, on the one hand, the conception of quantities with precise values, and on the other, what measurement can deliver. The difficulties are not simply a matter of limitations in precision, or evaluation of accuracy. They also derive from the inescapability of theory involvement in measurement. Ostensibly scientific descriptions refer to concrete entities and quantities in nature, but what are the referents if those descriptions can only be understood within their theoretical context? A naïve assumption of truth of the theory that supplies or constitutes the context might make this question moot, but that is exactly the attitude Teller takes out of play. Measurement is theory-laden, one might say, but laden with false theories! Teller argues that the main problems can be seen as an artifact of vagueness, and applies Eran Tal's robustness account of measurement accuracy to propose ways of dealing, with vagueness and idealization, to show that the theoretical problems faced by philosophical accounts of measurement are not debilitating to scientific practice.

In his commentary, van Fraassen insists that the identification of physical quantities is not a problem to be evaded. The question raised by Teller, how to identify the referent of a quantity term, is just the question posed in "formal mode" of what it means for a putative quantity to be real. But the way in which this sort of question appears in scientific practice does not answer to the sense it is given in metaphysics. It is rather a way of raising a question of adequacy for a scientific theory which concerns the extent to which values of quantities in its models are in principle determinable by procedures that the theory itself counts as measurements. On his proposal, referents of quantity terms are items in, or aspects of, theoretical models, and the question of adequacy of those models vis à vis data models replaces the metaphysical question of whether quantity terms have real referents in nature.

In his rejoinder, Paul Teller submits that van Fraassen's sortie to take metaphysics by the horns does not go far enough, the job needs to be completed. And that requires a pragmatist critique. As to that, Teller sees van Fraassen's interpretation of his, Teller's, critique of traditional measurement accuracy realism as colored by a constructive empiricist bias. Thus his rejoinder serves to do several things: to rebut the implied criticism in van Fraassen's comments and to give a larger-scale overview of the differences between the pragmatist and empiricist approaches to understanding science. New in this exchange is Teller's introduction of a notion of adoption of statements and theories, that has some kinship to van Fraassen's notion of acceptance of theories but is designed to characterize epistemic or doxastic attitudes that do not involve full belief at any point.

THE HISTORICAL AND METHODOLOGICAL CONTEXT

While experimentation and modeling were studied in philosophy of science throughout the twentieth century, their delicate entanglement and mutuality has recently come increasingly into focus. The essays in this collection concentrate on the experimental side of modeling, as well as, to be sure, the modeling side of experimentation.

In order to provide adequate background to these essays, we shall first outline some the historical development of this philosophical approach, and then present in very general terms a framework in which we understand this inquiry into scientific practice. This will be illustrated with case studies in which modeling and experimentation are saliently intertwined, to provide a touchstone for the discussions that follow.

A Brief History

Philosophical views of scientific representation through theories and models have changed radically over the past century.

EARLY TWENTIETH CENTURY: THE STRUCTURE OF SCIENTIFIC THEORIES

In the early twentieth century there was a rich and complex interplay between physicists, mathematicians, and philosophers stimulated by the revolutionary impact of quantum theory and the theory of relativity. Recent scholarship has illuminated this early development of the philosophy of science in interaction with avant-garde physics (Richardson 1997; Friedman 1999; Ryckman 2005) but also with revolutionary progress in logic and the foundations of mathematics.

After two decades of seminal work in the foundations of mathematics, including the epochal *Principia Mathematica* (1910–13) with Alfred North Whitehead, Bertrand Russell brought the technique of logical constructs to the analysis of physics in *Our Knowledge of the External World* (1914) and *The Analysis of Matter* (1927). Instants and spatial points, motion, and indeed the time and space of physics as well as their relations to concrete experience were subjected to re-creation by this technique. This project was continued by Rudolf Carnap in his famous *Der logische Aufbau der Welt* (1928) and was made continually more formal, more and more a part of the subject matter of mathematical logic and meta-mathematics. By midcentury, Carnap's view, centered on theories conceived of as sets of sentences in a formally structured language supplemented with relations to observation and measurement, was the framework within which philosophers discussed the sciences. Whether aptly or inaptly, this view was seen as the core of the logical positivist position initially developed in the Vienna Circle. But by this time it was also seen as contestable. In a phrase clearly signaling discontent, in the opening paragraphs of his "What Theories Are Not" (Putnam 1962), Hilary Putnam called it "the Received View."

It is not perhaps infrequent that a movement reaches its zenith after it has already been overtaken by new developments. We could see this as a case in point with Richard Montague's "Deterministic Theories" (Montague

1957), which we can use to illustrate both the strengths and limitations of this approach. Montague stays with the received view of theories as formulated in first-order predicate languages, but his work is rich enough to include a fair amount of mathematics. The vocabulary's basic expressions are divided, in the Carnapian way, into *abstract constants* (theoretical terms) and *elementary constants* (observational terms), with the latter presumed to have some specified connection to scientific observation procedures.

With this in hand, the language can provide us with sufficiently complete descriptions of possible trajectories ("histories") of a system; Montague can define: "A theory *T* is *deterministic* if any two histories that realize *T*, and are identical at a given time, are identical at all times. Second, a physical system (or its history) is *deterministic* exactly if its history realizes some deterministic theory." Although the languages considered are extensional, the discussion is clearly focused on the possible trajectories (in effect, alternative possible "worlds") that satisfy the theory. Montague announces novel results, such as a clear disconnection between periodicity and determinism, contrary to their intimate relationship as depicted in earlier literature.

But it is instructive to note how the result is proved. First of all, by this definition, a theory that is satisfied only by a single history is deterministic—vacuously, one might say—even if that history is clearly not periodic. Second, given any infinite cardinality for the language, there will be many more periodic systems than can be described by theories (axiomatizable sets of sentences) in that language, and so many of them will not be deterministic by the definition.

Disconcertingly, what we have here is not a result about science, in and by itself, so to speak, but a result that is due to defining determinism in terms of *what can be described in a particular language*.

MID-TWENTIETH CENTURY: FIRST FOCUS ON MODELS RATHER THAN THEORIES

Discontent with the traditional outlook took several forms that would have lasting impact on the coming decades, notably the turn to scientific realism by the Minnesota Center for the Philosophy of Science (Wilfrid Sellars and Grover Maxwell) and the turn to the history of science that began with Thomas Kuhn's *The Structure of Scientific Revolutions* (1962), which was first published as a volume in the International Encyclopedia of Unified Science. Whereas the logical positivist tradition had viewed scientific theoretical language as needing far-reaching interpretation to be understood, both these seminal developments involved viewing scientific language as part of natural language, understood prior to analysis.

But the explicit reaction that for understanding scientific representation the focus had to be on *models* rather than on a theory's linguistic formulation, and that models had to be studied independently as mathematical structures, came from a third camp: from Patrick Suppes, with the slogan "mathematics, not meta-mathematics!" (Suppes 1962, 1967).

Suppes provided guiding examples through his work on the foundations of psychology (specifically, learning theory) and on physics (specifically, classical and relativistic particle mechanics). In each case he followed a procedure typically found in contemporary mathematics, exemplified in the replacement of Euclid's axioms by the definition of Euclidean spaces as a class of structures. Thus, Suppes replaced Newton's laws, which had been explicated as axioms in a language, by the defining conditions on the set of structures that count as Newtonian systems of particle mechanics. To study Newton's theory is, in Suppes's view, to study this set of mathematical structures.

But Suppes was equally intent on refocusing the relationship between pure mathematics and empirical science, starting with his address to the 1960 International Congress on Logic, Methodology, and Philosophy of Science, "Models of Data" (Suppes 1962). In his discussion "Models versus Empirical Interpretations of Theories" (Suppes 1967), the new focus was on the relationship between theoretical models and models of experiments and of data gathered in experiments. As he presented the situation, "We cannot literally take a number in our hands and apply it to a physical object. What we can do is show that the structure of a set of phenomena under certain empirical operations is the same as the structure of some set of numbers under arithmetical operations and relations" (Suppes 2002, 4). Then he introduced, if still in an initial sketch form, the importance of data models and the hierarchy of modeling activities that both separate and link the theoretical models to the practice of empirical inquiry:

The concrete experience that scientists label an experiment cannot itself be connected to a theory in any complete sense. That experience must be put through a conceptual grinder . . . [Once the experience is passed through the grinder,] what emerges are the experimental data in canonical form. These canonical data constitute a model of the results of the experiment, and direct coordinating definitions are provided for this model rather than for a model of the theory . . . The assessment of the relation between the model of the

> experimental results and some designated model of the theory is a characteristic fundamental problem of modern statistical methodology. (Suppes 2002, 7)

While still in a "formal" mode—at least as compared with the writings of a Maxwell or a Sellars, let alone Kuhn—the subject has clearly moved away from preoccupation with formalism and logic to be much closer to the actual scientific practice of the time.

THE SEMANTIC APPROACH

What was not truly possible, or advisable, was to banish philosophical issues about language entirely from philosophy of science. Suppes offered a correction to the extremes of Carnap and Montague, but many issues, such as the character of physical law, of modalities, possibilities, counterfactuals, and the terms in which data may be presented, would remain. Thus, at the end of the sixties, a via media was begun by Frederick Suppe and Bas van Fraassen under the name "the Semantic Approach" (Suppe 1967, 1974, 2000; van Fraassen 1970, 1972).

The name was perhaps not quite felicitous; it might easily suggest either a return to meta-mathematics or alternatively a complete banishing of syntax from between the philosophers' heaven and earth. In actuality it presented a focus on models, understood independently of any linguistic formulation of the parent theory but associated with limited languages in which the relevant equations can be expressed to formulate relations among the parameters that characterize a target system.

In this approach the study of models remained closer to the mathematical practice found in the sciences than we saw in Suppes's set-theoretic formulations. Any scientist is thoroughly familiar with equations as a means of representation, and since Galois it has been common mathematical practice to study equations by studying their sets of solutions. When Tarski introduced his new concepts in the study of logic, he had actually begun with a commonplace in the sciences: *to understand an equation is to know its set of solutions.*

As an example, let us take the equation $x^2 + y^2 = 2$, which has four solutions in the integers, with x and y able to take either values +1 or −1. Reifying the solutions, we can take them to be the sequences <+1, +1>, <−1, +1>, <+1, −1>, and <−1, −1>. Tarski would generalize this and give it logical terminology: these sequences *satisfy* the sentence "$x^2 + y^2 = 2$." So when Tarski assigned sets of sequences of elements to sentences as their semantic values, he was following that mathematical practice of characterizing equations through their sets of solutions. It is in this fashion that one arrives at what in logical terminology is a *model*. It is a model *of* a certain set of equations if the sequences in the domain of integers, with the terms' values as specified, satisfy those equations. The set of all models of the equations, so understood, is precisely the set of solutions of those equations.[1]

The elements of a sequence that satisfy an equation may, of course, not be numbers; they may be vectors or tensors or scalar functions on a vector space, and so forth. Thus, the equation picks out a region in a space to which those elements belong—and that sort of space then becomes the object of study. In meta-mathematics this subject is found more abstractly: the models are relational structures, domains of elements with relations and operations defined on them. Except for its generality, this does not look unfamiliar to the scientist. A Hilbert space with a specific set of Hermitean operators, as a quantum-mechanical model, is an example of such a relational structure.

The effect of this approach to the relation between theories and models was to see the theoretical models of a theory as clustered in ways natural to a theory's applications. In the standard example of classical mechanics, the state of a particle is represented by three spatial and three momentum coordinates; the state of an N-particle system is thus represented by $3N$ spatial and $3N$ momentum coordinates. The space for which these $6N$-tuples are the points is the phase space common to all models of N-particle systems. A given special sort of system will be characterized by conditions on the admitted trajectories in this space. For example, a harmonic oscillator is a system defined by conditions on the total energy as a function of those coordinates. Generalizing on this, a theory is presented through the general character of a "logical space" or "state space," which unifies its theoretical models into families of models, as well as the data models to which the theoretical models are to be related, in specific ways.

REACTION: A CLASH OF ATTITUDES AND A DIFFERENT
CONCEPT OF MODELING

After the death of the Received View (to use Putnam's term), it was perhaps the semantic approach, introduced at the end of the 1960s, that became for a while the new orthodoxy, perhaps even until its roughly fiftieth anniversary (Halvorson 2012, 183). At the least it was brought into many areas of philosophical discussion about science, with applications extended, for example, to the philosophy of biology (e.g., Lloyd 1994). Thomas Kuhn exclaimed, in his outgoing address as president of the Philosophy of Science

Association, "With respect to the semantic view of theories, my position resembles that of M. Jourdain, Moliere's *bourgeois gentilhomme,* who discovered in middle years that he'd been speaking prose all his life" (1992, 3).

But a strong reaction had set in about midway in the 1980s, starting with Nancy Cartwright's distancing herself from anything approaching formalism in her *How the Laws of Physics Lie:* "I am concerned with a more general sense of the word 'model.' I think that a model—a specially prepared, usually fictional description of the system under study—is employed whenever a mathematical theory is applied to reality, and I use the word 'model' deliberately to suggest the failure of exact correspondence" (Cartwright 1983, 158–59). Using the term "simulacra" to indicate her view of the character and function of models, she insists both on the continuity with the semantic view and the very different orientation to understanding scientific practice:

> To have a theory of the ruby laser [for example], or of bonding in a benzene molecule, one must have models for those phenomena which tie them to descriptions in the mathematical theory. In short, on the simulacrum account the model is the theory of the phenomenon. This sounds very much like the semantic view of theories, developed by Suppes and Sneed and van Fraassen. But the emphasis is quite different. (Cartwright 1983, 159)

What that difference in emphasis leads to became clearer in her later writings, when Cartwright insisted that a theory does not just arrive "with a belly-full" of models. This provocative phrasing appeared in a joint paper in the mid-1990s, where the difference between the received view and the semantic approach was critically presented:

> [The received view] gives us a kind of homunculus image of model creation: Theories have a belly-full of tiny already-formed models buried within them. It takes only the midwife of deduction to bring them forth. On the semantic view, theories are just collections of models; this view offers then a modern Japanese-style automated version of the covering-law account that does away even with the midwife. (Cartwright, Shomar, and Suárez 1995, 139)

According to Cartwright and her collaborators, the models developed in application of a theory draw on much that is beside or exterior to that theory, and hence not among whatever the theory could have carried in its belly.

What is presented here is not a different account of what *sorts of things* models are, but rather a different view of the *role* of theories and their relations to models of specific phenomena in their domain of application. As Suárez (1999) put it, their slogan was that theories are not sets of models, they are *tools for the construction of models*. One type of model, at least, has the traditional task of providing accurate accounts of target phenomena; these they call *representative models*. They maintain, however, that we should not think of theories as in any sense containing the representative models that they spawn. Their main illustration is the London brothers' model of superconductivity. This model is grounded in classical electromagnetism, but that theory only provided tools for constructing the model and was not by itself able to provide the model. That is, it would not have been possible to just deduce the defining equations of the model in question after adding data concerning superconductivity to the theory.[2]

Examples of this are actually ubiquitous: a model of any concretely given phenomenon will represent specific features not covered in any general theory, features that are typically represented by means derived from other theories or from data.[3]

It is therefore important to see that the turn taken here, in the philosophical attention to scientific modeling in practice, is not a matter of logical dissonance but of approach or attitude, which directs how a philosopher's attention selects what is important to the understanding of that practice. From the earlier point of view, a model of a theory is a structure that realizes (satisfies) the equations of that theory, in addition to other constraints. Cartwright and colleagues do not present an account of models that contradicts this. The models constructed independently of a theory, to which Cartwright and colleagues direct our attention, do satisfy those equations—if they did not, then the constructed model's success would tend to refute the theory. The important point is instead that the *process of model construction* in practice was not touched on or illuminated in the earlier approaches. The important change on the philosophical scene that we find here, begun around 1990, is the attention to the fine structure of detail in scientific modeling practice that was not visible in the earlier more theoretical focus.[4]

TANGLED THREADS AND UNHERALDED CHANGES

A brief history of this sort may give the impression of a straightforward, linear development of the philosophy of science. That is misleading. Just a single strand can be followed here, guided by the need to locate the contributions in this volume in a fairly delineated context. Many other strands are entangled with it. We can regard David Lewis as continuing Carnap's program of the 1970s and 1980s (Lewis 1970, 1983; for critique see van Fraassen 1997). Equally, we can see Hans Halvorson as continuing in as well as correcting the semantic approach in the twenty-first century (Halvorson 2012, 2013; for discussion see van Fraassen 2014a). More intimately entangled with the attention to modeling and experimenting in practice are the writings newly focused on scientific representation (Suárez 1999, 2003; van Fraassen 2008, 2011). But we will leave aside those developments (as well as much else) to restrict this chapter to a proper introduction to the articles that follow. Although the philosophy of science community is by no means uniform in either focus or approach, the new attitude displayed by Cartwright did become prevalent in a segment of our discipline, and the development starting there will provide the context for much work being done today.

MODELS AS MEDIATORS

The redirection of attention to practice thus initiated by Cartwright and her collaborators in the early 1990s was systematically developed by Morgan and Morrison in their contributions to their influential collection *Models as Mediators* (1999). Emphasizing the autonomy and independence of theory, they describe the role or function of models as *mediating* between theory and phenomena.

What, precisely, is the meaning of this metaphor? Earlier literature typically assumed that in any scientific inquiry there is a background theory, of which the models constructed are, or are clearly taken to be, realizations. At the same time, the word "of" looks both ways, so to speak: those models are offered as representations *of* target phenomena, as well as being models *of* a theory in whose domain those phenomena fall. Thus, the *mediation* metaphor applies there: the model sits between theory and phenomenon, and the bidirectional "of" marks its middle place.

The metaphor takes on a stronger meaning with a new focus on how models may make their appearance, to represent experimental situations or target phenomena, before there is any clear application of, let alone derivation from, a theory. A mediator effects, and does not just instantiate, the relationship between theory and phenomenon. The model plays a role (1) in the representation of the target, but also (2) in measurements and experiments designed to find out more about the target, and then farther down the line (3) in the prediction and manipulation of the target's behavior.

Morrison and Morgan's account (Morrison 1999; Morrison and Morgan 1999) begins with the earlier emphasis on how model construction is not derivation from a theory but construction that draws on theory, on data, on other theories, on guiding metaphors, on a governing paradigm. That is the main content of the first thesis: *independence* in construction. The second thesis, *autonomy* of models, to the extent that it goes beyond this independence, can be illustrated by the delay, even negligence, with respect to the task of showing that the model proposed for a given phenomenon does actually satisfy the basic equations of the main theory.

For example, in fluid mechanics the basic principle in the background of all theorizing and model construction is the set of Navier–Stokes equations. There is no sense in which a specific fluid mechanics model, such as a model of turbulence in the wake of an obstacle, was ever deduced from those equations alone. But there is so little suspicion that a given model, proposed in practice, violates those equations that eventual checking on the consistency is left to mathematicians, without the experimenter waiting for reassurance. Morrison illustrates the point with Ludwig Prandtl's 1904 construction of a model of the boundary layer model for viscous fluids. His construction employs the tools of classical hydrodynamics and the Navier–Stokes equations, but "the important point is that the approximations used in the solutions come not from a direct simplification of the mathematics of the theory, but from the phenomenology of the fluid flow as represented in the model" (Morrison, 1999, 59).

That models are autonomous and independent in this sense does not by itself reveal the character of the role of mediation. The term "mediator" connotes a bridging of some sort between two disparate sides in a dialogue, dispute, or collaboration. So it is crucial to appreciate the two sides: while theories are drawn on to construct models, conversely the models aid in theory construction. In particular cases, a model may come along first, the extension of theory following upon what was learned while a resistant phenomenon was being modeled. That models function as mediators between theory and the phenomena implies then that modeling can enter in two ways. The process of modeling may start with a phenomenon (physical system or process) and draw on theory to devise a representation of that phenomenon; or it may start with a theory, draw on other sources such as data or auxiliary theories to compliment that theory, and introduce as model a structure satisfying that theory. In the first case it is (or is intended to be) an accurate representation of a phenomenon; in the second case it is a representation of what the theory depicts as going on in phenomena of this sort.

As is typical of metaphors, drawing out the content of "models mediate between theory and phenomena" turns out to be a complex but instructive exercise.

RECENT DEVELOPMENTS

As in science, so in philosophy: practice goes far beyond what is preached. By the second decade of the twenty-first century, in which the present collection is situated, the previous developments had borne ample fruit. In two new series of conferences, *Models and Simulations* and *The Society for Philosophy of Science in Practice*, starting, respectively, in France in 2006 and in the Netherlands in 2007, the new orientation to scientific practice has been saliently displayed. Of signal importance in the work by Nancy Cartwright, Mary Morgan, Margaret Morrison, Mauricio Suárez, and the participants in these conferences was the detailed examination of actual scientific practice in experimentation and modeling.

More or less concurrently with the workshops of 2009–2011, a bevy of new studies appeared on how computer simulation was changing conceptions of modeling, measurement, and experiment. These included works by our contributors Ronald Giere (2009), Isabelle Peschard (2011b, 2012, 2013), Michael Weisberg (2013), Eric Winsberg (2009, 2010), and by, for example, Anouk Barberousse, Sara Franceschelli, and Cyrille Imbert (2009), Paul Humphreys (2004), Margaret Morrison (2009), E. C. Parke (2014), and Wendy Parker (2009). The main issue about computer simulation, evoking sustained debate in the literature, was saliently expressed by Margaret Morrison: Do computer simulations ever have the same epistemic status as experimental measurement? Morrison had argued earlier (as we have seen) that models function, in some sense, as measuring instruments; she now argued that there is a way in which simulation can be said to constitute an experimental activity (see further Morrison 2015).

At the same time, as debated in the workshops, there were new arguments concerning the theory-dependence of measurement. This surfaces in the present collection especially in the contributions by Joseph Rouse and Paul Teller, but such discussion continued in a series of publications subsequently: Ann-Sophie Barwich and Hasok Chang (2015), Nancy Cartwright

(2014), Teru Miyake (2013, 2015), Eran Tal (2011, 2012, 2013), and Bas van Fraassen (2012, 2014b).

At issue here are the roles of idealization, abstraction, and prediction in establishing measurement outcomes as well as the theoretical status of criteria to determine what counts as a measurement and what it is that is measured. A special issue of *Studies in the History and Philosophy of Science, Part A* (vol. 65–66, October–December 2017) is dedicated to the history and philosophy of measurement in relation to modeling and experimentation.

Experimentation and Modeling: A Delicate Entanglement

In the remainder of this introduction we will sketch in broad outlines, with illustrative case studies, a framework in which we see the experimental side of modeling currently approached.[5]

As the philosophical take on modeling changed, so did views of how experimentation relates to modeling. The earlier stages inherited the logical positivist view of experimentation as the *tribunal* that tests models against data delivered by experimental and observational outcomes. In later stages much attention was paid to how measurement itself involves modeling from the outset.[6]

There is, as might be expected, another side to the coin as well: that, conversely, experimental activity makes a constructive contribution to the processes of modeling and simulating. The autonomy of modeling was a new theme that is to be conjoined with another new theme: the constructive relation of experimentation to modeling as an interactive, creative, open-ended process that modifies both along the way.

There are two aspects to this interaction. The first is that the specification of the relevant parameters of a phenomenon are not given from the outset. The phenomenon vaguely defined at the beginning of the investigation needs to be specified, and this is done by specifying what data represent the phenomenon and what data are to be regarded as a manifestation of the phenomenon. What needs to be settled through experimentation will then include the conditions in which measurement outcomes qualify as relevant data—that is, data to which a putative model of the phenomenon is accountable. The second, which follows upon this, is conceptual innovation, as a result of an effort to make sense, through modeling, of the phenomenon and the conditions of its occurrence.

Looking back from our present vantage point we may discern illustrative examples in the past: surely it was experimentation that led to the reconceptualization of lightning as electric discharge, for example. But the intricate ballet between experimental and modeling progress can only become clear through a detailed analysis, and for this we shall choose a case in fluid mechanics.

EXAMPLE OF A PHENOMENON: FORMATION OF A WAKE

The formation of a wake is a very common phenomenon that happens when air or liquid goes over a bluff (not streamlined) body, which can be a pole, a rock, or an island. The work on a better theoretical understanding of wakes spread from meteorology to the stability of bridges or platforms, from the design of cars and airplane wings to that of helicopter vanes—and more generally to all cases where periodic instabilities or transitions toward chaotic behavior are possible.

A simple physical model of this phenomenon in a laboratory can show the wake of a flow behind a cylinder when the velocity of the upstream flow reaches a certain critical value. In a diagram such a flow going, say, from left to right can be visualized in the plane perpendicular to the axis of the cylinder, with the wake formed by vortices that are emitted alternatively on each side of the cylinder and carried away with the downstream flow.

As simple as it may look, the attempt to construct a theoretical model of this sort of wake triggered an enormous number of studies, and no less controversy. As our main example from this literature let us take *one*, seemingly simple, question that was the object of a debate involving experimental as well as numerical and theoretical studies in fluid mechanics in the second half of the twentieth century.

Formulation of such a question begins inevitably within a pre-existing modeling tradition. The system is initially characterized in terms of three quantities: the velocity (U) upstream of the flow, the diameter (d) of the cylinder, and the viscosity (v) of the fluid. The most significant quantity, defined in terms of these three, is the dimensionless Reynolds number (Re):

$$Re = Ud/v$$

The wake is formed when this number reaches a critical value, where vortices are emitted with a certain frequency, the *shedding frequency*.

> Question: What happens when Re is increased within a certain interval beyond the critical value?

How does the evolution of the shedding frequency of the vortices vary with Re? Specifically, as Re is increased within a certain interval beyond the critical value *is the variation with the Re of the shedding frequency a continuous linear variation or is there some discontinuity?*

The question can itself arise only within a theoretical background, but it clearly asks for data from experiment before models of the formation of the wake can be assessed and also, in effect, before a suitable model of the entire phenomenon can be constructed.

ORIGIN OF THE CONTROVERSY: THE DISCONTINUITY

A detailed experimental study of the wake appeared in 1954 with the publication of Anatol Roshko's dissertation "On the Development of Turbulent Wakes from Vortex Streets." The experimental results showed that in the range (40–150) of Re "regular vortex sheets are formed and no turbulent motion is developed." This is called the stable range. Between Re = 150 and Re = 300 turbulent velocity fluctuations accompany the periodic formation of vortices: this is the range of turbulence. For the stable range Roshko provided an empirical formula for the increase of shedding frequency of velocity: a linear variation of the shedding frequency with the Reynolds number.

But in a new study Tritton (1959) called into question that there were only two ranges of shedding, the stable and the turbulent, and directly contradicted Roshko's results regarding the evolution of the shedding frequency with the Reynolds number. Tritton argued on the basis of new measurements for the existence of a discontinuity in the curve that displays the frequency plotted against the velocity. This *discontinuity* appears within the stable range, thus contradicting the linear relationship of Roshko's formula.

In addition, Roshko's simple division into two ranges of shedding, one stable the other turbulent, suggested that the dynamics of the wake in the stable range would be two-dimensional, contained in the plane perpendicular to the cylinder. Tritton's visualization of the wake appeared to show, to the contrary, that the dynamics of the wake are not what Roshko's results suggested. Beyond the discontinuity—that is, for values of Re greater than the one for which the discontinuity occurs—the shedding of the vortices along the cylinder is not simultaneous. To put it differently, the imaginary lines joining side-by-side vortices along the cylinder are not parallel to the axis of the cylinder—they are oblique.

That the successive lines of the vortices are or are not parallel to the axis of the cylinder translates in terms of the dimension of the dynamics of the wake. Parallel lines of vortices correspond to a two-dimensional dynamics of the wake. In contrast, nonparallel lines of vortices testify to the existence of a dynamics in the direction of the cylinder, which added to the two-dimensional dynamics would make the total dynamics of the wake three-dimensional. But three-dimensional effects on the dynamics were thought to be associated with the development of turbulence, which according to Roshko took place *beyond* the stable range.

This conflict between Roshko's and Tritton's experimental results started a controversy that lasted thirty years. Is or is not the discontinuity, and the oblique shedding, an intrinsic, fluid-mechanic phenomenon, irrespective of the experimental setup?

MODEL IMPLICATIONS VERSUS EXPERIMENTAL MEASUREMENTS

It was not until 1984 that a model of the wake was proposed to account for its temporal dynamics—that is, for the temporal evolution of the amplitude of the vortices and of the frequency at which they are emitted (Mathis, Provansal, and Boyer 1984). The model in question was obtained from the general model proposed by Landau (1944) to describe the development of a periodic instability, which he viewed as the first step toward turbulence. As illustrated in Figure I.1, the Landau model of the wake describes the amplitude of the wake in the two-dimensional plane perpendicular to the axis of the cylinder, and it predicts that the maximum amplitude is proportional to the difference between the Reynolds number (Re) and its critical value (Re$_c$):

$$U^2 y_{\max} \propto (\mathrm{Re} - \mathrm{Re}_c)$$

The measurements of the amplitude that were made showed that, in this respect, the model works beautifully—even better than expected. So for the evolution of the amplitude, at least, one and the same model can account for the development of the instability on the whole range of the Reynolds number. This result contradicts Tritton's claim that two different instabilities are at play on two ranges of Reynolds number.

But the same model also predicts that the evolution of the frequency with the Reynolds number is linear, with no discontinuity! Yet the measurement results continue to show the existence of a discontinuity. And additional measurements made along the cylinder indicate the existence of a three-dimensional dynamics, an oblique shedding.

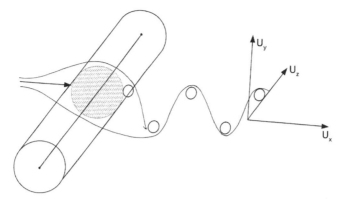

Figure I.1. Two-dimensional geometry: cross-section of the cylinder. (Courtesy P. Le Gal. Adapted by Nousoul Digital Publishers.)

ANALYSIS: BACK TO QUESTIONS OF INTERPRETATION

Does the discrepancy between the model's prediction of the evolution of the frequency and the outcomes of measurement show or even indicate that the Landau model is not an adequate model for the wake? That depends. It depends on whether that discontinuity has to be accounted for by a model of the wake. If the discontinuity is an artifact, the model of the wake not only does not have to account for it, but should not account for it. On the other hand, if it is an intrinsic feature of the wake, a model that does not account for it cannot, in that context, count as a model of the wake.

The problem is not one of data analysis or of construction of what is usually referred to, after Suppes (1962), as "models of the data." The main problem can be posed apart from that. Even if procedures of analysis are assumed to be in place and a data model is produced—a data model of the evolution of the shedding frequency with Re—we are still left with an open question: Is this data model one that the model of the wake *should* match?

This has the form of a normative question. How is it settled in practice? Since it was in fact settled in practice, the subsequent history provides us with an instructive lesson in how fact and normativity interact in scientific inquiry.

INTRINSIC CHARACTERISTICS AND RELEVANT PARAMETERS

Williamson (1989) described the controversy as a search to determine *"whether the discontinuity is an intrinsic, fluid-mechanic phenomenon, irrespective of the experimental setup."* The idea of being irrespective of the experimental setup seems to offer an empirical criterion to distinguish genuine data—data that are informative about the target phenomenon—from uninformative data, including artifacts. If the discontinuity is intrinsic, it should not depend on the experimental setup; if it is shown to depend on the experimental setup, then it is not intrinsic. This motivated experimental studies of the effect of an increase of the non-uniformities in the flow or in the diameter, as well as of the effect of making a cylinder vibrate. In each case, the idea was to show that the discontinuity is generated by some specific features of the experimental setup and consequently is not a feature of the wake itself.

It is not sufficient, however, to show that the discontinuity is only the effect of non-uniformities or vibrations. It would also have to be shown that without non-uniformities or vibrations there is no discontinuity. This is precisely the challenge that some numerical studies were going to try to address.

SIMULATION OF THE WAKE

It is not easy to show that when there is *no* non-uniformity or *no* vibration there is no discontinuity. Both flowing fluid and the diameter of a cylinder keep a certain level of non-uniformity, however carefully they are prepared. Fortunately, by the end of the 1980s the situation of the wake lent itself to the modern alternative: numerical simulation.

A simulation of the Navier–Stokes equations, which are fundamental equations in fluid mechanics, was performed to find out how the flow behind a cylinder develops when there are no non-uniformities of any sort and no vibration (Karniadakis and Triantafyllou 1989). The results of the simulation were presented as pointing to a definite answer to the question of the nature of the discontinuity. And once again, as with the Landau model, the answer was that the evolution of the frequency with Re is linear, with no discontinuity.

These results certainly show that the occurrence of the discontinuity in the experiments results from the influence of some "additional" factors that are not taken into account as parameters of the system in the Navier–Stokes equations, as applied to this setup in the simulation. But the parameters of Navier–Stokes are those whose effect is constitutive of fluid-mechanical phenomena. So if one trusts the method used for the simulation (a spectral-element method, used successfully in previous studies) and does not envisage calling into question the validity of the fundamental equations, the most obvious conclusion would be that the effect of these additional factors constitutes an artifact and should therefore be shielded.

This conclusion about the discontinuity only holds, however, under certain assumptions. For the results to be relevant to the understanding of this phenomenon, the simulation must be an imitation, an accurate mimetic representation of the phenomenon we are interested in. Whether it is indeed, this is where the problem of identification of the relevant parameters sneaks in.

To speak of simulating the fundamental equations is not exactly right in at least two respects. First of all, the computer can only run a discrete model. Thus, the simulation requires the construction of a system of discrete equations and a method of discretization for time and space to obtain the simulation model. As Lenhard (2007) shows, the construction of the simulation model may become a modeling process in its own right, when the norm that guides and regulates the construction is precisely in agreement with the observations of the phenomenon in question. The main normative requirement being the successful imitation, the simulation model may be as far from the theoretical model as from a phenomenological model.

Something else may be overlooked when one speaks of simulating the fundamental equations, something that is independent of the way in which the simulation model is obtained. The fundamental equations are abstract, and going from the fundamental equations to the simulation of a target phenomenon must involve specifications that determine what particular situation is the target of the simulation.

On closer inspection, what raises doubt as to the significance of the result of the simulation is the geometry of the simulated situation. As may already have been apparent from Figure 1.1, this is a two-dimensional geometry representing a plane containing a cross section of the cylinder.

The simulation is meant to tell what the observation should be, what the phenomenon is really like, and whether the discontinuity is part of it or not. But how could this simulation of the development of a flow in a two-dimensional plane tell what it is like when a flow goes around a cylinder and develops in a space that contains not only the plane perpendicular to the axis of the cylinder but also the plane that contains that axis?

There is an assumption that answers this question. It is the assumption that with respect to the phenomenon under study (the frequency of shedding of the vortices forming the wake) all points on the cylinder are in relevant respects equivalent to one another, that "the same thing happens everywhere." With that assumption in place, there is no need to simulate the wake in each point of the cylinder; any cross section will suffice.

What the two-dimensional simulation shows then is how the wake develops, according to the Navier–Stokes equations, in conditions where all the points on the cylinder are relevantly interchangeable. But why should we think that all the points are interchangeable? The presence of the ends obviously creates an asymmetry contradicting the assumptions of the simulation.

To this question too there is an answer. Suppose that a cylinder that is long enough can be regarded as infinite, as a cylinder that has no end. If there is no end, then we are in the situation where all points are interchangeable. All that is needed to satisfy this assumption of an infinite cylinder is that, for a long enough cylinder, what happens in the middle part of the cylinder be independent from what happens at or near the ends. And it could then be admitted that the two-dimensional simulation will, at least, show what should happen in a long enough cylinder, far enough from the ends.

Taking the simulation as relevant is, consequently, taking the ends of the cylinder as being irrelevant to the understanding of the fluid-mechanical features, amplitude or frequency, of the wake. Another way to put this: so regarded, the ends of the cylinder are treated in the same way as non-uniformities of the flow or vibrations of the cylinder. If they have an effect on the outcomes of measurement, this effect will be classified as an artifact and should be shielded.

Thus, the implicitly made assumption is that the ends are taken not to be a relevant parameter of the system, and that the effects on the dynamics of the wake that are due to a finite cylinder having ends are not intrinsic characteristics of the dynamics.

EXPERIMENTAL CONTRIBUTION TO CONCEPTUAL UNDERSTANDING

This assumption about the ends of the cylinder would be temporarily supported by measurements that had shown that for a long enough cylinder the frequency of shedding in the middle of the cylinder is different from that found near the ends. But that should not mislead us into thinking that the assumption was an empirical assumption. The assumption is normative, in that it specifies the "normal" conditions of development of the wake, the conditions where it has its "pure" form. With this in place, the conditions under which the ends would have an effect on the measurement results would simply not count as the proper conditions of measurement.

There was for the experimenter an additional assumption in place: that the difference between a finite cylinder and one with no ends depends just on the length of the cylinder. Concretely, this implies the assumption that the way to shield off the effect of the ends is to have a sufficient length.

These two assumptions were called into question by Williamson (1989) in a thoroughgoing experimental study of the evolution of the shedding frequency, which was a turning point for our understanding of the discontinuity and the development of three-dimensional effects.

Measurements of the shedding frequency with a probe moving along the span of the cylinder showed the existence of different regions characterized by different shedding frequencies. In particular, a region of lower frequency was found near the ends. More precisely, for a cylinder with an aspect ratio (the ratio L/D of the length to the diameter) beyond a specific value, the frequency near the ends differed from the frequency in the central region. "This suggests," Williamson wrote, "that the vortex shedding in the central regions of the span is unaffected by the *direct* influence from the end conditions" (Williamson 1989, 590; italics added). Note, however, that Williamson recognized only the absence of a *direct* influence.

Why did Williamson underline the absence only of a *direct* influence of the ends on the wake in the central region? In the case where there was a difference in frequency between the ends and the central part of the cylinder, visualizations of the temporal development of the wake along the cylinder were made. They showed that, initially, the lines of vortices traveling downstream were parallel to the cylinder and that progressively the parallel pattern was transformed into a stable oblique pattern, which propagated from the ends of the cylinder toward the central region. These observations suggested that there is an effect propagated toward the center. If so, this could be attributed to an influence of the ends, which is *indirect* in that it is not on the value of the frequency itself. But whether this influence should be part of our understanding of the wake would still be a question.

So far, all the observations and measurements had been made with endplates *perpendicular* to the axis of the cylinder. But with this new focus on the possibility of effects due to the ends, further measurements were made for different values of the angle between the axis of the cylinder and the plates. And, lo and behold, for a certain angle the shedding becomes parallel—that is, two-dimensional—and the discontinuity disappears, even though *the length did not change.*

Changing the angle of the plates has the effect of changing the *pressure* conditions responsible for the existence of a region of lower frequency toward the ends. When there is such a region of lower frequency, a phase difference propagates from the ends toward the central region, and this propagation creates the pattern of oblique shedding. For a certain interval of angles of the endplates, when the pressure and the vortex frequency match those values over the rest of the span, there is no region of lower frequency and no propagation of phase difference, and the shedding is parallel. The infamous discontinuity only appears in the oblique mode of shedding and is found to correspond to the transition of one oblique pattern to another with a slightly different geometry.

ANALYSIS: AN INTERACTIVE, CREATIVE, OPEN-ENDED PROCESS

Williamson takes his results to "show that [the oblique and parallel patterns] are both intrinsic and are simply solutions to different problems, because the boundary conditions are different" (1989, 579). The two forms of shedding correspond to different values of the angle between the endplates and the axis of the cylinder. If no special status is bestowed on certain values of this angle in contrast to the others, there is no reason to take only one of the shedding patterns as being normal or intrinsic. In this new perspective, the parallel and the oblique pattern are not two distinct phenomena, with only one being the normal form of the wake. They are two possible configurations of the flow corresponding to different values of a parameter of the experimental system, two possible solutions for the same system in different conditions.

But this new way of seeing implies that the two assumptions, on which the relevance of the simulation depended, must be rejected. First, a new parameter should be added to the set of relevant parameters of the system, namely, one that characterizes the end conditions of the cylinder. This is to insist that the phenomenon under study is a process involving a *finite* cylinder because exactly finite cylinders are ones with ends; the effect that the end conditions have on the development of the wake is now part of the structural characteristics of the wake. Second, this parameter is independent of the length of the cylinder. The difference between the ends and the central part needs to be reconceived in terms of pressure difference and the value of the angle of the end plates that determines the value of this pressure difference.

By integrating this parameter among the set of relevant parameters the gain is one of conceptual unification: what were seen as two distinct phenomena have been unified under the same description. To integrate the ends among the relevant factors through the definition of a new relevant parameter and not to bestow a special status on a particular range of values of the angle are normative transformations of the investigation.

To sum up, the elaboration of an experimental system is an interactive, creative, open-ended process and contributes constructively to the processes of modeling and simulating. The constructive contribution is mediated by the identification of the relevant parameters. The relevant parameters are characteristics of the experimental system such that not only does their variation have an effect on the phenomenon but this effect is constitutive of the phenomenon—intrinsic to the phenomenon. As we have seen, the classification of characteristics into intrinsic versus interference or artifact is not there beforehand; it is during the inquiry, with its successive steps of experiment and model construction, that the phenomenon is identified. To paraphrase a line from a quite different philosophical scene, *the phenomenon under study is what it will have been:* what was studied is what it is seen to have been in retrospect.

The identification of the relevant parameters is required for determining the conditions in which measurements provide the empirical touchstone of a putative model of the phenomenon. Before that, a putative model is untestable. The specification of the relevant parameters involves a systematic empirical investigation of the effects of different factors, but the line that is drawn between which effects are relevant and which are not is normative. The effects of the relevant parameters are those a model of the phenomenon should account for. The criterion of relevance and the consequent criteria of adequacy for modeling determine the normative "should."

We had an interactive process in that both the predictions of a still untestable model and the results of a prejudiced simulation contributed to shaping the experimental search for the relevant parameters. The new relevant parameter that was introduced in the conception of the phenomenon amounted to a conceptual innovation. The process of mutually entangled steps of experimentation and modeling is a creative process.

And it is open ended. A new model was formulated in response to the reconception of the phenomenon. Immediately, the exactitude of some of Williamson's measurements was called into question on the basis of an analysis of the solutions of that new model. New measurements were to follow as well as new simulations and a modified version of the model, and so it goes on.[7]

Models of Experiments and Models of Data

An experiment is a physical, tangible realization of a *data-generating procedure,* designed to furnish information about the phenomena to which a theoretical model or hypothesis pertains. But while it is correct that the experiment and the procedure performed are physical and tangible, it would be thoroughly misleading to regard them merely as thus. The experimenter is working with a *model* of the instrumental setup, constructed following the general theoretical models afforded in the theoretical background.

MODEL OF THE EXPERIMENT

In Pierre Duhem's *The Aim and Structure of Physical Theory* (1914/1962) we find our first inspiring effort to describe the interactive practice that constitutes the experimental side of modeling. Duhem describes graphically the synoptic vision required of the experimenting scientist:

> When a physicist does an experiment, two very distinct representations of the instrument on which he is working fill his mind: one is the image of the concrete instrument that he manipulates in reality; the other is a schematic model of the same instrument, constructed with the aid of symbols supplied by theories; and it is on this ideal and symbolic instrument that he does his reasoning, and it is to it that he applies the laws and formulas of physics. (153–54)

This is then illustrated with Regnault's experiment on the compressibility of gases:

> For example, the word manometer designated two essentially distinct but inseparable things for Regnault: on the one hand, a series of glass tubes, solidly connected to one another, supported on the walls of the tower of the Lycée Henri IV, and filled with a very heavy metallic liquid called mercury by the chemists; on the other hand, a column of that creature of reason called a perfect fluid in mechanics, and having at each point a certain density and temperature defined by a certain equation of compressibility and expansion. It was on the first of these two manometers that Regnault's laboratory assistant directed the eyepiece of his cathetometer, but it was to the second that the great physicist applied the laws of hydrostatics. (156–57)

We would say it somewhat differently today, if only to place less emphasis on the instrument than on the setup as a whole. But Duhem's insight is clear: the scientist works both with the material experimental arrangement and, inseparably, indissolubly, with a model of it closely related to, though generally not simply derived from, a theoretical model.

EXPERIMENT AS DATA-GENERATING PROCEDURE

For the modeling of phenomena, what needs to be explored and clarified is the experimental process through which theoretical models or hypotheses about phenomena get "connected" to the world through the acquisition of data. Several steps in this process need to be distinguished and analyzed.

As Patrick Suppes made salient meanwhile, it is not just the model of the experimental setup that appears in the experiment's progress (Suppes 1962, 2002). The experiment is a data-generating procedure, and the experimenter is constructing a *data model:* "The concrete experience that scientists label an experiment cannot itself be connected to a theory in any complete sense. That experience must be put through a conceptual grinder . . . [Once the experience is passed through the grinder,] what emerges are the experimental data in canonical form" (Suppes 2002, 7). Thus, we see alongside the model of the experimental setup and the model of the phenomenon a third level of modeling: the modeling of the data. But as Suppes emphasized, even that is an abstraction because many levels within levels can be distinguished.

We find ourselves today a good century beyond Duhem's introduction to the modeling of experiments and a good half-century beyond Suppes's spotlight on the data model. In those seminal texts we see a conceptual development that was far from over. Unexpected complexities came to light in both aspects of the interactive process of experimentation and modeling.

CONSTRUCTION OF THE DATA MODEL, THEN AND NOW

To show how data models are constructed we will select two examples from experimental work in physics, approximately one hundred and twenty years apart, so as to highlight the commonality of features in modern experimental work.[8] The first example will be Albert A. Michelson's 1879 experiment at the Naval Academy to determine the velocity of light. As second illustration we have chosen a contemporary article in fluid mechanics, Cross and Le Gal's 2002 experiment to study the transition to turbulence in a fluid confined between a stationary and a rotating disk. In both these experiments, the reported results consist of the data generated, presented in (quite differently displayed) summary "smoothed" form but with ample details concerning the raw data from which the data model was produced.

Michelson on Determination of the Velocity of Light

The terrestrial determination of the velocity of light was one of the great experimental problems of the nineteenth century. Two measurement results were obtained by Fizeau's toothed wheel method, by Fizeau in 1849 and by Cornu in 1872. Between these, Foucault had obtained a result in 1862 that was of special interest to Michelson.

A schematic textbook description of Foucault's apparatus sounds simple: a beam of light is reflected off a rotating mirror R to a stationary mirror M, whence it returns to R, and there it is reflected again in a slightly different direction because R has rotated a bit meanwhile. The deflection (the angle between the original and new direction of the beam) is a function of three factors: the distance between the mirrors, the rotational velocity of R, and the speed of light. The first two being directly measurable, the latter can be calculated.

But this omits an important detail. There is a lens placed between R and M to produce an image of the light source on M. That image moves across M as R rotates, and if the distance is to be increased (so as to increase the deflection), mirror M must be made larger. Foucault managed to increase the distance to about sixty-five feet, but that left the deflection still so small that the imprecision in its measurement was significant.

Michelson realized that by placing the light source at the principal focus of the lens he would produce the light beam as a parallel bundle of rays, and that could be reflected back as a parallel beam by a plane mirror placed at any desired distance to the rotating mirror. Even his first setup allowed for a distance of 500 feet, with a deflection about twenty times larger than in Foucault's. The experiment in 1879 improved significantly on that, and the value he obtained was within 0.0005 km/second of the one we have today. This value is what is reported in Michelson (1880).

So what were the data generated on which this calculation was based? They are presented in tables, with both raw data and their means presented explicitly. Michelson is careful to present their credentials as unbiased observation results, when he displays as a specimen the results noted on June 17, 1879, at sunset: "the readings were all taken by another and noted down without divulging them till the whole five sets were completed." We note that, at the same time, this remark reveals the human limits of data collection at that time, which will be far surpassed in the contemporary example we present later. The columns in Table I.1 are sets of readings of the micrometer for the deflected image. Following this specimen comes the summary of all results from June 5 through July 2, 1879; Table I.2 is a truncated version of just the first two days.

Concentrating on this part of the report omits many of the details and the ancillary reported observations, including those that serve to certify the

Table I.1. Sets of readings of the micrometer for the deflected image (Michelson 1880)

	112.81	112.80	112.83	112.74	112.79
	81	81	81	76	78
	79	78	78	74	74
	80	75	74	76	74
	79	77	74	76	77
	82	79	72	78	81
	82	73	76	78	77
	76	78	81	79	75
	83	79	74	83	82
	73	73	76	78	82
Mean	112.801	112.773	112.769	112.772	112.779

Table I.2. Results from June 5 through June 7, 1879 (Michelson 1880)

Date	Distinctness of image	Temperature (°F)	Position of deflected image	Difference between greatest and lowest values	Number of revolutions per second	Radius (feet)	Velocity of light in air (km)
5	3	76	114.85	0.17	257.36	28.672	299,850
7	2	72	114.64	0.10	257.52	28.655	299,740
7	2	72	114.58	0.08	257.52	28.647	299,900
7	2	72	85.91	0.12	193.14	28.647	300,070
7	2	72	85.97	0.07	193.14	28.650	299,930
7	2	72	114.61	0.07	257.42	28.650	299,850

calibration and reliability of the instruments used. For the steel tape used to measure distances, the micrometer, and the rate of the tuning fork for time measurement, the report includes a separate data-generating procedure and its results. The *credentialing* of the experimental setup concludes with a detailed consideration of possible sources of error, such as the possibility that the rotation of the mirror could throw the deflected light in the direction of rotation, or that the rotation of the mirror itself could cause a distortion by twisting or centrifugal force, or that there could be systematic bias in a single observer's readings. Although just a few paragraphs are devoted to the theory (in effect, the previous schematic description), what is described in detail is the procedure by which the data are generated and the care and precautions to ensure reliability, issued in a summary presentation of those data and the calculated value of the speed of light.

Cross and Le Gal on Torsional Couette Flow

As so much else in fluid mechanics, the flow of a fluid between two plates, with one rotating relative to the other (Couette flow), can be modeled starting with the Navier–Stokes equations. But the study of idealized models—for example, with two infinite parallel plates, even with a pressure gradient imposed—leaves the experimenter with only clues as to what to expect in real situations of this sort.

In the study by Cross and Le Gal (2002), the subject of study was turbulence in the flow for different values of the rotation rate Ω. The gap h between the two disks could be varied continuously between 0 and 21 mm, and the rotational velocity between 0 and 200 revolutions per minute (rpm). The specific results included in the report, as presenting a typical case, were for a fixed gap of approximately 2.2 mm, with the rotation rate Ω increased by 2 rpm in steps from 42 rpm to 74 rpm.

The report of the experimental results shows the same attention to the details of the apparatus and setup that we saw in Michelson's report, but there is a world of difference in the techniques whereby the data model is produced—as we are more than a century beyond hand-drawn tables of numbers. Figure I.2 shows the experimental device, which consists of a water-filled cylindrical housing in which the rotating disk is immersed and whose top is the stationary disk. The raw data are obtained by use of a video camera. The water is seeded with reflective anisotropic flakes to perform visualizations. The orientation of these particles depends upon the shear stress of the flow, so the structures that develop in the fluid layer can be observed. For each Ω value, the velocity of the rotating camera is chosen so that the turbulent spirals are seen as stationary to facilitate the image acquisition.

These images become the raw material for the production of the data model. The turbulent regions appear dark because of the completely disordered motion of the reflective flakes in these areas. The obtained spatiotemporal diagrams are thus composed of turbulent/dark domains inside laminar/clear ones. To extract physical characteristics, the diagrams are binarized: the disordered states appear in black, the ordered in white. The filtered images are produced from the raw images by sharpening the images, followed by a succession of "erosions" and "dilatations."[9]

This procedure so far already involves a good deal of processing of the data, but it is still an intermediate step on the way to the data model that depicts the significant results. Upon increasing the rotating disk velocity Ω,

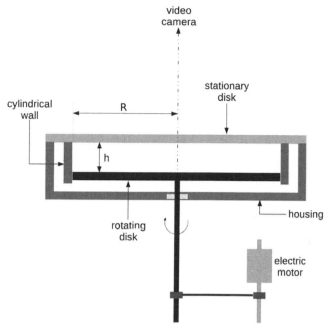

Figure I.2. Experimental setup for Couette flow. (Courtesy P. Le Gal. Adapted by Nousoul Digital Publishers.)

a first instability results that leads to a periodic spiral wave pattern. When Ω is increased still further, this wave pattern presents some amplitude modulations and phase defects. At a second threshold, these defects trigger the turbulent spirals. As the rotation rate is increased, the lifetime of these turbulent structures increases until a threshold is reached where they then form permanent turbulent spirals arranged nearly periodically all around a circumference. However, because the number of these turbulent spirals decreases with the rotational frequency, the transition to a fully turbulent regime is not achieved. The data model thus described, constructed from a series of experiments of the type described here, is graphically and rather dramatically presented in Figure I.3.

ANALYSIS: FROM SINGULAR OBSERVATION STATEMENT
TO GIGABYTES OF DATA

The difference in the procedures followed, at a concrete level, to manage large amounts of data collected not by individual observation but application of technology is clear. In this respect Michelson's procedure is already very

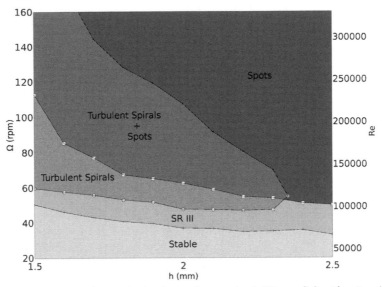

Figure I.3. Transition diagram plotting the rotation rate threshold in rpm (left axis) against the gap width to show the turbulent spirals' appearance threshold (curves marked with o) and disappearance threshold. (Courtesy P. Le Gal. Adapted by Nousoul Digital Publishers.)

sophisticated, but Cross and Le Gal's is quantitatively far beyond Michelson's with respect to the generation, collection, recording, and statistical analysis of the data. To appreciate what is important about these two examples of experimentation, we need to appreciate their function, which is essentially the same but must be discerned within these very different technological realizations.

The data model constructed on the basis of outcomes of measurement can be viewed as the pivotal element in the experimental process. What makes the data model a pivotal element is its dual epistemic function:

- On the one hand, the data model is supposed to be an empirical representation of the phenomenon under investigation.
- On the other hand, the data model serves as a benchmark for the construction and evaluation of the theoretical model of (or theoretical hypothesis about) the phenomenon.

Given the pivotal function of the data model, to investigate the experimental process that connects theoretical models to phenomena is to clarify the different steps that lead to the construction and interpretation of data models:

- The formulation of a model of the experiment, specifying what needs to be measured (motivated and guided by a preconception of the phenomenon under investigation and, in some cases, a putative theoretical model of this phenomenon).
- The realization and interpretation of measurement operations according to the specifications of the model of the experiment.
- The construction of data models on the basis of the outcomes of measurement.
- The interpretation of the data models, as repository of information about phenomena.
- The use of data models to construct or produce evidence for/against the theoretical model offered as putative theoretical representation of the phenomenon.

The arrows in Figure I.4 do not represent temporal order. To put it bluntly, even paradoxically, experimentally based modeling starts neither with a model to be tested nor with a data-generating procedure. It starts with a problem for which relevant data need to be produced and evidentially

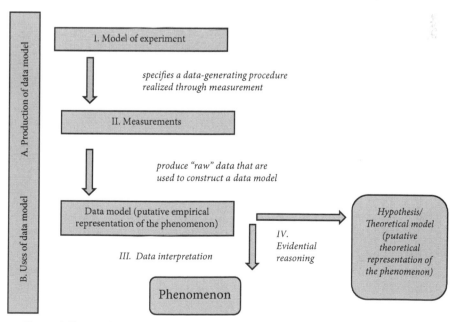

Figure I.4. The experimental process diagrammatically displayed.

supported models need to be constructed. The different stages in this process, from the abstract terms of the problem to the specification and realization of experimental procedures, from issues about evidential relevance to an explication of the similarity relation between models and data, are the subject of the contributions in this collection.

THE EXPERIMENTAL SIDE OF MODELING
PHILOSOPHICALLY APPROACHED

The philosophical essays on modeling and experimentation in this volume are representative of the current literature in the philosophy of science in many ways: in their diversity of approach, in the tensions between them, and in the differences in focus that divide them. One crucial and central concern is found in all of them—the most striking, groundbreaking result perceivable in these contributions is their radically new conception of the role of data in the modeling process. Specifically, they have a new awareness of what is problematic about data. Instead of being a passive element of modeling or experimenting, already standing ready to act the role of tribunal or simply flowing from an experimental procedure, data have become the central point of the experimentally based modeling process. Data are both what need to be produced and what need to be accounted for to create the prospect for articulation of both the theoretical model and the model of the experiment. Data are the manifestation of the phenomenon under study, they are what needs to be interpreted, and their very relevance or significance itself becomes the issue.

NOTES

1. Logical terminology is typically foreign to the background mathematical practice: a set of models that is precisely thus, the set of models that satisfy a given set of sentences, is in logical terminology an *elementary class*. But let us not let differences of jargon obscure that the logicians' "model" is just a straightforwardly generalized concept coming from ordinary mathematical practice familiar to the scientist.

2. Reflecting on her journey from 1983 to 1995, in a later article Cartwright wrote, "I have been impressed at the ways we can put together what we know from quantum mechanics with much else we know to draw conclusions that are no part of the theory in the deductive sense" (1999, 243). The point is, of course, not that one cannot deduce all consequences of a conjunction from one conjunct—obviously—but rather to draw attention to

the cooperation with other input so saliently, importantly needed in actual practice.

3. Superconductivity presented an instructive example. A much earlier example in the literature that should already have made the point glaringly clear was Jon Dorling's 1971 analysis of how Einstein arrived at the photon model of light.

4. See also appendix 1 of van Fraassen (2008).

5. For a more comprehensive version of this section, see Peschard (2010).

6. Indeed, to Margaret Morrison (2009) it did not seem inapt to say that models, including those models that are simulation programs on computers, can function as measuring instruments. But this was certainly more controversial, especially for computer simulations. There is definitely an understanding of models as having a role in experimenting and measuring that does not make them measuring instruments (cf. Peschard 2013; Parke 2014).

7. For the larger story, see further Peschard (2011a).

8. The term "data model" as we currently use it in the philosophy of science derives from the seminal writings by Patrick Suppes. This usage is to be distinguished from its use in other fields such as software engineering in database construction for business and industry.

9. The erosion is the elimination of a black pixel if a chosen number s or more of white pixels is found among its eight neighbors. A dilatation is the inverse transformation: a white pixel is transformed to black if it is surrounded by a number s' or more of black pixels. Here, s, s' were set at 2. The first step is an erosion in order to eliminate the undesirable black dots from noise. Then seven successive dilatations and a final erosion achieve a visual aspect of the binarized diagrams relevantly equivalent to the original ones.

REFERENCES

Barberousse, Anouk, Sara Franceschelli, and Cyrille Imbert. 2009. "Computer Simulations as Experiments." *Synthese* 169: 557–74.

Barwich, Ann-Sophie, and Hasok Chang. 2015. "Sensory Measurements: Coordination and Standardization." *Biological Theory* 10: 200–211.

Bogen, James, and James Woodward. 1988. "Saving the Phenomena." *Philosophical Review* 98: 303–52.

Carnap, Rudolf. 1928. *Der logische Aufbau der Welt.* Berlin-Schlachtensee: Welt-kreis Verlag.

Cartwright, Nancy. 1979. "Causal Laws and Effective Strategies." *Noûs* 13 (4): 419–37.

Cartwright, Nancy D. 1983. *How the Laws of Physics Lie.* Oxford: Clarendon Press.

Cartwright, Nancy D. 1999. "Models and the Limits of Theory: Quantum Hamiltonians and the BCS Models of Superconductivity." In *Models as Mediators: Perspectives on Natural and Social Science,* edited by Mary S. Morgan and Margaret Morrison, 241–81. Cambridge: Cambridge University Press.

Cartwright, Nancy D. 2014. "Measurement." In *Philosophy of Social Science. A New Introduction,* edited by N. Cartwright and E. Montuschi, 265–87. New York: Oxford University Press.

Cartwright, Nancy D., T. Shomar, and M. Suárez. 1995. "The Tool Box of Science." In *Theories and Models in Scientific Processes,* edited by W. Herfel, W. Krajewski, I. Niiniluoto, and R. Wojcicki, 137–49. Amsterdam: Rodopi.

Cross, Anne, and Patrice Le Gal. 2002. "Spatiotemporal Intermittency in the Torsional Couette Flow between a Rotating and a Stationary Disk." *Physics of Fluids* 14: 3755–65.

Dorling, Jon. 1971. "Einstein's Introduction of Photons: Argument by Analogy or Deduction from the Phenomena?" *British Journal for the Philosophy of Science* 22: 1–8.

Duhem, Pierre. (1914) 1962. *The Aim and Structure of Physical Theory.* Translated by Philip P. Wiener. New York: Atheneum. First published as *La Théorie physique, son objet, sa structure* (Paris: Marcel Rivière & Cie, 1914).

Friedman, Michael. 1999. *Reconsidering Logical Positivism.* Cambridge: Cambridge University Press.

Giere, Ronald N. 2009. "Is Computer Simulation Changing the Face of Experimentation?" *Philosophical Studies* 143: 59–62.

Halvorson, Hans. 2012. "What Scientific Theories Could Not Be." *Philosophy of Science* 79: 183–206.

Halvorson, Hans. 2013. "The Semantic View, if Plausible, Is Syntactic." *Philosophy of Science* 80: 475–78.

Humphreys, Paul. 2004. *Extending Ourselves: Computational Science, Empiricism, and Scientific Method.* Oxford: Oxford University Press.

Karniadakis, G. E., and G. S. Triantafyllou. 1989. "Frequency Selection and Asymptotic States in Laminar Wakes." *Journal of Fluid Mechanics* 199: 441–69.

Knuuttila, Tarja, and Andrea Loettgers. 2014. "Varieties of Noise: Analogical Reasoning in Synthetic Biology." *Studies in History and Philosophy of Science Part A* 48: 76–88.

Kuhn, Thomas S. 1962. *The Structure of Scientific Revolutions.* International Encyclopedia of Unified Science, vol. 2, no. 2. Chicago: University of Chicago Press.

Kuhn, Thomas S. 1992. "Presidential Address: Introduction." In *PSA: Proceedings of the Biennial Meeting of the Philosophy of Science Association* 1992 (2): 3–5.

Landau, L. 1944. "On the Problem of Turbulence." *Comptes Rendus Académie des Sciences U.S.S.R.* 44: 311–14.

Lenhard, Johannes. 2007. "Computer Simulation: The Cooperation between Experimenting and Modeling." *Philosophy of Science* 74: 176–94.

Lewis, David K. 1970. "How to Define Theoretical Terms." *Journal of Philosophy* 67: 427–46.

Lewis, David K. 1983. "New Work for a Theory of Universals." *Australasian Journal of Philosophy* 61: 343-377.

Lloyd, Elisabeth A. 1994. *The Structure and Confirmation of Evolutionary Theory.* 2nd ed. Princeton, N.J.: Princeton University Press.

Mathis, C., M. Provansal, and L. Boyer. 1984. "The Benard-Von Karman Instability: An Experimental Study near the Threshold." *Journal de Physique Lettres* 45: L483–91.

Michelson, Albert A. 1880. "Experimental Determination of the Velocity of Light: Made at the U.S. Naval Academy, Annapolis." In *Astronomical Papers of the U.S. Nautical Almanac* 1, Part 3, 115–45. Washington, D.C.: Nautical Almanac Office, Bureau of Navigation, Navy Department. http://www.gutenberg.org/files/11753/11753-h/11753-h.htm.

Miyake, Teru. 2013. "Underdetermination, Black Boxes, and Measurement." *Philosophy of Science* 80: 697–708.

Miyake, Teru. 2015. "Reference Models: Using Models to Turn Data into Evidence." *Philosophy of Science* 82: 822–32.

Montague, Richard. 1957. "Deterministic Theories." In *Decisions, Values and Groups,* edited by D. Wilner and N. F. Washburne, 325–70. New York: Pergamon. Reprinted in *Formal Philosophy: Selected Papers of Richard Montague,* edited by Richmond Thomason (New Haven, Conn.: Yale University Press, 1974).

Morgan, Mary, and Margaret Morrison, eds. 1999. *Models as Mediators: Perspectives on Natural and Social Science.* Cambridge: Cambridge University Press.

Morrison, Margaret. 1999. "Models as Autonomous Agents." In *Models as Mediators: Perspectives on Natural and Social Science,* edited by Mary S. Morgan and Margaret Morrison, 38–65. Cambridge: Cambridge University Press.

Morrison, Margaret. 2009. "Models, Measurement and Computer Simulations: The Changing Face of Experimentation." *Philosophical Studies* 143: 33–57.

Morrison, Margaret. 2015. *Reconstructing Reality: Models, Mathematics, and Simulations.* Oxford: Oxford University Press.

Morrison, Margaret, and Mary S. Morgan. 1999. "Models as Mediating Instruments." In *Models as Mediators: Perspectives on Natural and Social Science,* edited by Mary S. Morgan and Margaret Morrison, 10–37. Cambridge: Cambridge University Press.

Parke, E. C. 2014. "Experiments, Simulations, and Epistemic Privilege." *Philosophy of Science* 81: 516–36.

Parker, Wendy S. 2009. "Does Matter Really Matter? Computer Simulations, Experiments and Materiality." *Synthese* 169: 483–96.

Peschard, Isabelle. 2010. "Target Systems, Phenomena and the Problem of Relevance." *Modern Schoolman* 87: 267–84.

Peschard, Isabelle. 2011a. "Making Sense of Modeling: Beyond Representation." *European Journal for Philosophy of Science* 1: 335–52.

Peschard, Isabelle. 2011b. "Modeling and Experimenting." In *Models, Simulations, and Representations,* edited by P. Humphreys and C. Imbert, 42–61. New York: Routledge.

Peschard, Isabelle. 2012. "Forging Model/World Relations: Relevance and Reliability." *Philosophy of Science* 79: 749–60.

Peschard, Isabelle. 2013. "Les Simulations sont-elles de réels substituts de l'expérience?" In *Modéliser & simuler: Epistémologies et pratiques de la modélisation et de la simulation,* edited by Franck Varenne and Marc Siberstein, 145–70. Paris: Editions Materiologiques.

Putnam, Hilary. 1962. "What Theories Are Not." In *Logic, Methodology and Philosophy of Science: Proceedings of the 1960 International Congress,* edited by E. Nagel, P. Suppes, and A. Tarski, 240–51. Stanford, Conn.: Stanford University Press. Reprinted in *Mathematics, Matter and Method. Philosophical Papers,* 215–27, vol. 1 (Cambridge: Cambridge University Press, 1979).

Richardson, Alan. 1997. *Carnap's Construction of the World: The Aufbau and the Emergence of Logical Empiricism.* Cambridge: Cambridge University Press.

Roshko, Anatol. 1954. *On the Development of Turbulent Wakes from Vortex Streets.* Washington, D.C.: National Advisory Committee for Aeronautics. http://resolver.caltech.edu/CaltechAUTHORS:ROSnacarpt1191

Russell, Bertrand. 1914. *Our Knowledge of the External World: A Field for Scientific Method in Philosophy.* Chicago: Open Court.

Russell, Bertrand. 1917. "Mysticism and Logic." In *Mysticism and Logic and Other Essays,* 1–32. London: George Allen & Unwin. https://www.gutenberg.org/files/25447/25447-h/25447-h.htm

Russell, Bertrand. 1927. *The Analysis of Matter.* London: Allen & Unwin.

Ryckman T. A. 2005. *The Reign of Relativity: Philosophy in Physics 1915–1925.* New York: Oxford University Press.

Suárez, Mauricio. 1999. "Theories, Models and Representations." In *Model-Based Reasoning in Scientific Discovery,* edited by L. Magnani, N. Nersessian, and P. Thagard, 75–83. Dordrecht, the Netherlands: Kluwer Academic.

Suárez, Mauricio. 2003. "Scientific Representation: Against Similarity and Isomorphism." *International Studies in the Philosophy of Science* 17: 225–44.

Suppe, Frederick Roy. 1967. *On the Meaning and Use of Models in Mathematics and the Exact Sciences,* PhD diss., University of Michigan.

Suppe, Frederick Roy. 1974. *The Structure of Scientific Theories.* Urbana: University of Illinois Press.

Suppe, Frederick Roy. 2000. "Understanding Scientific Theories: An Assessment of Developments, 1969–1998." *Philosophy of Science* 67 (Proceedings): S102–15.

Suppes, Patrick. 1962. "Models of Data." In *Logic, Methodology and Philosophy of Science: Proceedings of the 1960 International Conference,* edited by E. Nagel, P. Suppes, and A. Tarski, 252–61. Stanford, Calif.: Stanford University Press.

Suppes, Patrick. 1967. *Set-Theoretical Structures in Science.* Stanford: Institute for Mathematical Studies in the Social Sciences, Stanford University.

Suppes, Patrick. 2002. "Introduction." In *Representation and Invariance of Scientific Structures,* 1–15. Stanford, Calif.: Center for the Study of Language and Information. http://web.stanford.edu/group/cslipublications/cslipublicationspdf/1575863332.rissbook.pdf

Tal, Eran. 2011. "How Accurate Is the Standard Second?" *Philosophy of Science* 78: 1082–96.

Tal, Eran. 2012. *The Epistemology of Measurement: A Model-Based Account,* PhD diss., University of Toronto. http://hdl.handle.net/1807/34936

Tal, Eran. 2013. "Old and New Problems in Philosophy of Measurement." *Philosophy Compass* 8: 1159–73.

Tritton, D. J. 1959. "Experiments on the Flow Past a Circular Cylinder at Low Reynolds Numbers." *Journal of Fluid Mechanics* 6: 547–67.

Van Fraassen, Bas C. 1970. "On the Extension of Beth's Semantics of Physical Theories." *Philosophy of Science* 37: 325–34.

Van Fraassen, Bas C. 1972. "A Formal Approach to the Philosophy of Science." In *Paradigms and Paradoxes: The Challenge of the Quantum Domain,* edited by R. Colodny, 303–66. Pittsburgh: University of Pittsburgh Press.

Van Fraassen, Bas C. 2008. *Scientific Representation: Paradoxes of Perspective.* Oxford: Oxford University Press.

Van Fraassen, Bas C. 2011. "A Long Journey from Pragmatics to Pragmatics: Response to Bueno, Ladyman, and Suarez." *Metascience* 20: 417–42.

Van Fraassen, Bas C. 2012. "Modeling and Measurement: The Criterion of Empirical Grounding." *Philosophy of Science* 79: 773–84.

Van Fraassen, Bas C. 2014a. "One or Two Gentle Remarks about Halvorson's Critique of the Semantic View." *Philosophy of Science* 81: 276–83.

Van Fraassen, Bas C. 2014b. "The Criterion of Empirical Grounding in the Sciences." In *Bas van Fraassen's Approach to Representation and Models in Science,* edited by W. J. Gonzalez, 79–100. Dordrecht, the Netherlands: Springer Verlag.

Weisberg, Michael. 2013. *Simulation and Similarity: Using Models to Understand the World.* New York: Oxford University Press.

Whitehead, Alfred North, and Bertrand Russell. 1910–13. *Principia Mathematica,* 3 vols. Cambridge: Cambridge University Press.

Williamson, C. H. K. 1989. "Oblique and Parallel Modes of Vortex Shedding in the Wake of a Circular Cylinder at Low Reynolds Number." *Journal of Fluid Mechanics* 206: 579–627.

Winsberg, Eric. 2009. "Computer Simulation and the Philosophy of Science." *Philosophy Compass* 4/5: 835–45.

Winsberg, Eric. 2010. *Science in the Age of Computer Simulation.* Chicago: University of Chicago Press.

MODELS OF EXPERIMENTS

RONALD N. GIERE

In the logical empiricist philosophy of science, the primary epistemological relationship was between a theory, T, and an observation, O. Both were understood as some sort of linguistic entities—sentences, statements, propositions, whatever. With the increasing emphasis on models within the philosophy of science, this relationship has increasingly been replaced by one between theoretical models and models of data. Moreover, following Suppes's (1962) long ago advice, it has been emphasized that there is a hierarchy of theoretical models, ranging from very general to very specific, between high level theoretical models and models of data. During roughly the same period as models came to the fore within the philosophy of science, some philosophers of science, together with sociologists and historians of science, independently developed an interest in experimentation. Only belatedly has it been realized that perhaps in some place in the hierarchy of models there should be a *model of the experiment*. This discussion explores that possibility.

In fact, the expression "model of the experiment" appears already in Suppes's 1962 article. It turns out to be something fairly abstract. Imagine an experiment (Suppes's example was taken from behaviorist learning theory) with a binary experimental variable and just two possible results on a particular trial. There are then four possible states for each trial, thus 4^n possible different series of n trials. Very roughly, Suppes identifies "the theory of the experiment" with this set of possible series. A model of the experiment is then a particular series that might be the realization of an actual experiment. What I have in mind is something far less abstract.

Closer to the mark is what Allan Franklin called a "theory of the apparatus" in his early work on the philosophy of experimentation (1986, 168). His idea was that a "well-confirmed theory of the apparatus" may give us reason to believe the results of experiments using that apparatus. I expect many others have subsequently invoked a similar idea. It is surely implicit in Ian Hacking's founding philosophical work on experimentation (1983). But concentrating on an apparatus seems still too narrow a focus for an account of a model of an experiment. If we are to incorporate experimentation into a model-based philosophy of science, we will need something broader.

A HIERARCHY OF MODELS

Figure 1.1 is my attempt to create a Suppes-like hierarchy of models incorporating a model of the experiment. It is clear that, in any particular case, the model of the experiment must be at the bottom of the hierarchy. The nature of the experiment determines what one might call the "form" of the data. That is, the experiment determines what kind of data will be acquired and, in the case of numerical data, the range of possible values of any particular data point. But before saying more about models of experiments, a word about the rest of the hierarchy is required.

As I understand this hierarchy, the principled models at the top are defined by what are usually called the "laws" of the science in question, such as Newton's laws or Maxwell's equations. The laws define a type of system instantiated abstractly in the principled models (a "theoretical perspective"). Adding further conditions, such as force functions in the case of classical mechanics, yields representational models. Representational models apply to designated real systems, or types of systems, with the features of these systems interpreted and identified in terms that define the principled models. Eventually this hierarchy reaches a fully specified representational model of a particular system involved in an experiment. It is this model that is compared with a model of the data generated by an experiment in order to decide how well the representational model fits the real system under investigation.

This picture presumes a unified mature science that includes high-level principles that can be fitted out for different circumstances. It is possible to construct hierarchies for cases in which multiple principles from different sciences are involved, or in which there are no general principles at all and thus no principled models.[1]

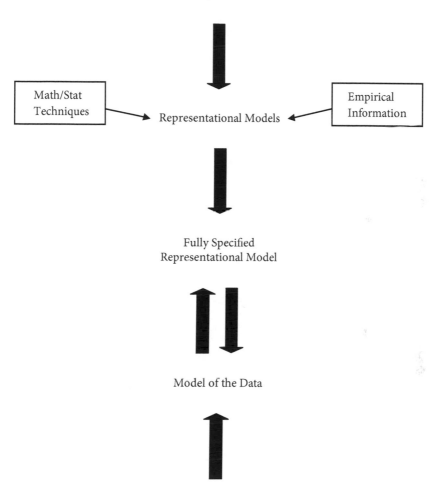

Figure 1.1. A Suppes-like hierarchy of models incorporating a model of the experiment.

In recognition of the messiness of applying abstract theoretical models (e.g., Morgan and Morrison 1999), I have included reference to the various mathematical and statistical techniques involved in making the many approximations and idealizations typically involved in eventually reaching a model of a specific real system. Also included is reference to the fact that aspects of models from other disciplines as well as empirical generalizations may need to be incorporated in the process. The details of this process are represented by the light horizontal arrows. The top two thick arrows pointing downward include deduction, but much else besides. The thick arrow going up from the model of the experiment represents the constraints the experiment places on any possible model of the data. Finally, up and down arrows between the model of the data and the fully specified representational model represent a comparison, and perhaps mutual adjustment, of these two models.

I think it would be a mistake to "flatten" the representation of experimenting with models to the extent that one loses any sense of an underlying hierarchical structure. Keeping the hierarchy turns out to be very useful in locating a model of the experiment in our understanding of experimenting with models.

Figure 1.1 realizes Suppes's idea of science as involving a hierarchy of models. Necessarily missing from the figure, however, is any representation of interaction with the world. Thus, although there is a model of the data, the data are not represented in the hierarchy. Models are entities, typically abstract, humanly constructed for many purposes, but prominently for representing aspects of the world. The data are one such aspect. To fill out the picture, I think we need a parallel, but separate, hierarchy, a physical rather than symbolic hierarchy. Figure 1.2 shows one possibility. It is the recorded data, of course, that are used to construct the model of the data in the Suppes hierarchy. Here the vertical arrows represent physical processes, but, of course, the data have to be "interpreted" in terms of the model of the data.

Recorded "Data"

The Physical Experiment

The World

Figure 1.2. A physical rather than symbolic hierarchy.

MODELS OF EXPERIMENTS

To proceed further it will help to introduce an example. Here I will invoke a contemporary experiment that I have explored in some detail elsewhere, observations with instruments aboard the satellite based Compton Gamma Ray Observatory.[2] The general theory that motivates this particular experiment is that heavy elements are produced inside very large stars. One such is a radioactive isotope of aluminum that decays with a distinctive emission of a 1.8 MeV Gamma Ray. The nearest example of large star formation is the center of our Milky Way galaxy. Thus, the observation of 1.8 MeV Gamma Rays from this source would provide direct confirmation of the theory of heavy element formation in stars. Figure 1.3 shows a Suppes hierarchy for this example.

Now we can begin to consider what a model of this experiment might be like. Here I will distinguish three categories of components that I think must be represented in any model of an experiment of this scope: material, computational, and agential (done by human agents).

Material Components

Material components include what in more traditional experiments are called "the apparatus." The primary material component in this case is the Compton Observatory itself, which includes four different detectors. I will focus on just one, the Compton Gamma Ray Telescope, or COMPTEL. As its name indicates, this instrument detects gamma rays by means of the Compton effect, the interaction of a gamma ray with an electron, which, moving through an appropriate substance, gives off light whose intensity can be measured and used to compute the energy of the incoming gamma ray. Using two levels of detectors, one can determine the angle at which the gamma ray entered the instrument, thus locating its source in the heavens. This requires complex coincidence–anticoincidence circuitry to ensure that it is the same gamma ray being detected twice, the second time with diminished energy.

The point here is that the knowledge required to develop a model of the instrument (just part of "the experiment") is different from that to which the experiment is meant to contribute—in this case, models of higher element formation in stars. So although I have located the model of the experiment at the bottom of a Suppes hierarchy (see Figure 1.1), the construction of this model takes one far outside the initial hierarchy. In fact, we can construct a completely different hierarchy of models leading to a model of the instrument, as shown in Figure 1.4.

In this case, the references to mathematical techniques and empirical information are mere markers for a world of engineering models and techniques beyond the ken of most philosophers of science.[3] The downward

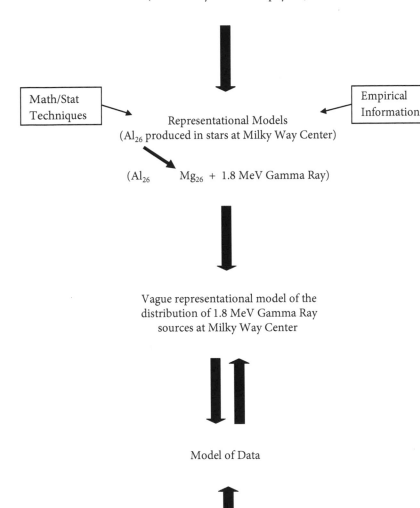

Figure 1.3. Suppes hierarchy for heavy element formation in stars.

arrows represent a correspondingly complex process leading finally to a model of COMPTEL. Note, finally, that this hierarchy is just one component of a comprehensive model of the experiment.

Computational Components

The whole experiment incorporates a vast computational network extending from the initial data acquisition to the final output as a computer graphic. Moreover, this network is vastly extended in both space and time. The telescope itself is, of course, in Earth orbit. Then there are relay satellites transmitting data down to earth stations in, for example, White Sands, New Mexico. Finally the data are analyzed at the Space Telescope Science Institute in Greenbelt, Maryland. On a temporal scale, some experiments include data acquired over months, even years. A model of this network, at least in schematic form, must be part of the final model of the experiment. And here again, one is appealing to knowledge completely outside the original Suppes hierarchy.

Agential Components

It obviously takes a great number of people to run a modern experiment as well as complex organization and careful planning. Perhaps the most salient role for agents is in the interpretation of the results, the comparison of the data model with relevant theoretical models. My own preference is to regard this not as any sort of logical inference but as individual and collective decision making—deciding what to conclude about the relative fit of theoretical models with the world, and with what degree of confidence.[4]

Overall, I would regard the operation of an experiment as a case of distributed cognition and the whole experimental setup as a distributed cognitive system.[5] The important point here, however, is that, once again, constructing a model of the experiment requires drawing on concepts and knowledge well beyond the original Suppes hierarchy.

UNDERSTANDING THE DATA MODEL

It has always been true that understanding the data from any experiment requires knowledge of the process that produced it. The framework of connecting a data model with a model of the experiment gives prominence to this fact and provides a vocabulary for thinking about it. In the case of COMPTEL, perhaps the most important scientific fact about the process is

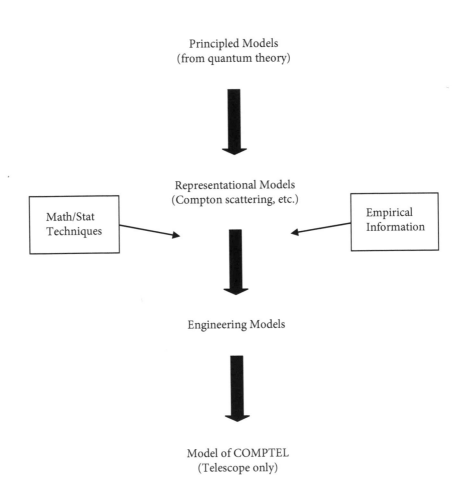

Figure 1.4. A hierarchy of models leading to a model of the instrument.

that it is based on Compton scattering. The rest is engineering, but of course it is the engineering that produces the accuracy in measurements of the energy of gamma rays and the location of the sources. The primary lesson is that understanding a data model requires understanding a model of the experiment.

The most salient output from a COMPTEL experiment is a computer graphic, together with information as to how it is to be read (e.g., the values of a variable associated with color coding). Which is to say, the output is already a model of the data. What would be considered the "raw" data is information about individual gamma ray interactions. This information is recorded, of course, and can be accessed to answer specific questions. But the graphics programming already has built into it algorithms that convert large numbers of individual "hits" into a continuous graphic. Now that philosophers of science have become aware of the importance of distinguishing between data and a model of the data, the actual process of producing a model of the data from the original data has largely become hidden in the design of the hardware and software that are part of the experiment.

Many contemporary experiments, like those involving satellite-based observatories, yield very large data sets—thousands and even millions of individual data points. The resulting data models are thus incredibly rich; they are so rich that that they can serve as good substitutes for the lowest level theoretical models. In fact, in many instances we do not know enough about the initial conditions to produce a bottom-level theoretical model that could be expected to fit the real system as well as our best data model. This was the case with knowledge of the distribution of gamma ray sources in the center of the Milky Way. It was expected on theoretical grounds that there would be sources of 1.8 MeV gamma rays, but theory, even supplemented with optical examination of the center of the Milky Way, provides little detailed knowledge of the distribution of those sources. The data model thus provides the best representational model we have of this distribution.

THE MAIN CONCLUSIONS are (1) understanding a data model requires having a good model of the experiment, and (2) producing a model of an experiment requires going outside the context of the traditional Suppes hierarchy of models. In particular, it requires consideration of the material, computational, and agential aspects of the experiment. Figure 1.5 is an attempt to capture these conclusions in a single diagram. The horizontal double arrow represents the interpretation of the data in terms of the concepts

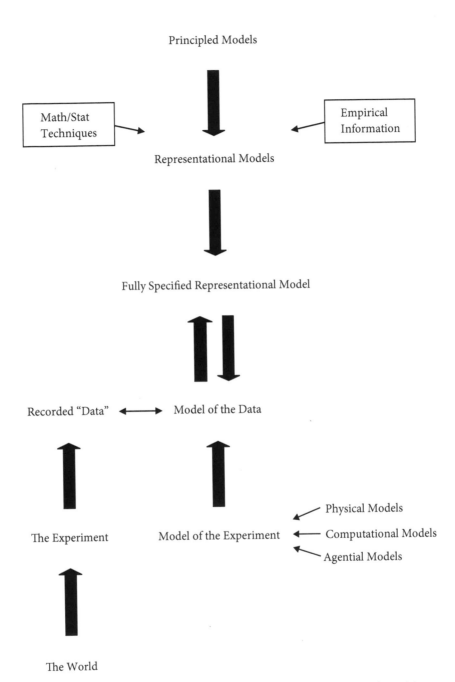

Figure 1.5. A model of an experiment that considers the material, computational, and agential aspects.

of the model of the data and the subsequent incorporation of the data into the data model. Finally, the example of the Compton Gamma Ray Telescope suggested several noteworthy features present in many contemporary experiments. (1) The operative output of many experiments is not data, but a data model. The construction of a data model is built into the hardware, with some of the software presumably modifiable as desired. (2) In "data rich" contexts, a data model may function as a representational model. It may effectively be the richest representational model available.

NOTES

1. I discuss these other possible hierarchies in Giere (2010).

2. For more information about this example, including color reproductions of some data, see Giere (2006), chapter 3.

3. A major exception is Nancy Cartwright (1999), and her earlier works.

4. I elaborated this view of scientific judgment in Giere (1985), reprinted with minor changes in Giere (1999). See also Giere (1988), chapter 6.

5. This view of scientific cognition is elaborated in Giere (2006), chapter 5.

REFERENCES

Cartwright, Nancy D. 1999. *The Dappled World: A Study of the Boundaries of Science.* Cambridge: Cambridge University Press.

Franklin, Allan. 1986. *The Neglect of Experiment.* Cambridge: Cambridge University Press.

Giere, Ronald N. 1985. "Constructive Realism." In *Images of Science,* edited by P. M. Churchland and C. A. Hooker, 75–98. Chicago: University of Chicago Press.

Giere, Ronald N. 1988. *Explaining Science: A Cognitive Approach.* Chicago: University of Chicago Press.

Giere, Ronald N. 1999. *Science without Laws.* Chicago: University of Chicago Press.

Giere, Ronald N. 2006. *Scientific Perspectivism.* Chicago: University of Chicago Press.

Giere, Ronald N. 2010. "An Agent-Based Conception of Models and Scientific Representation." *Synthese* 172 (2): 269–81.

Hacking, Ian. 1983. *Representing and Intervening.* Cambridge: Cambridge University Press.

Morgan, Mary, and Margaret Morrison, eds. 1999. *Models as Mediators: Perspectives on Natural and Social Science.* Cambridge: Cambridge University Press.

Suppes, Patrick. 1962. "Models of Data." In *Logic, Methodology and Philosophy of Science: Proceedings of the 1960 International Conference,* edited by E. Nagel, P. Suppes, and A. Tarski, 252–61. Stanford, Calif.: Stanford University Press.

2

DYNAMICS, DATA, AND NOISE IN
THE COGNITIVE SCIENCES

ANTHONY CHEMERO

SEMI-NAÏVE EXPERIMENTAL DESIGN

I have taught introductory laboratory psychology many times over the years. The psychology department at Franklin and Marshall College, where I taught this course regularly, is nearly unique in its commitment to teaching psychology as a laboratory science, and it is one of the very few institutions in the United States in which the first course in psychology is a laboratory course. It is a valuable experience for students, especially those who do not go on to major in a science, as it gives them a rudimentary understanding of hypothesis testing, experimental design, and statistics. It also reinforces for them that psychology really is an experimental science, having nothing to do with the fantasies of a famous Viennese physician. (Unfortunately, they mostly remember the rats.) The first meeting of the course stresses the very basic concepts of experimental design, and the lecture notes for that meeting include this sentence:

> The goal of a good experiment is to maximize primary variance, eliminate secondary variance, and minimize error variance. (Franklin and Marshall Psychology Department 2012, 3)

This sentence, which I will henceforth call the *methodological truism,* requires some unpacking. Experiments in psychology look at the relationship between an independent variable (IV) and a dependent variable (DV). The experimenter manipulates the IV and looks for changes in measurements of the DV. Primary variance is systematic variance in the DV caused by the manipulations of the IV. Secondary variance is systematic variance in the

DV caused by anything other than the manipulations of the IV. Error variance is unsystematic variance in the DV, or noise. The methodological truism, then, is the key principle of experimental design. A good experiment isolates a relationship between the IV and DV, and eliminates effects from other factors. Of course, although unsystematic variance in the DV is inevitable in our noisy world, it should be minimized to the extent possible. Because noise is inevitable, psychologists and other scientists tend to use some measure of the central tendency of the collection of measurements of the DV, such as the mean. This, in effect, washes out the unsystematic variance in the DV, separating the data from the noise.

Of course, the psychology department at Franklin and Marshall is not alone in teaching this material this way. You can find a version of the methodological truism in the textbook used for any introductory course in experimental methods. One of the claims I will argue for here is that the methodological truism is not, in general, true. I will argue for this point by describing experiments that use unsystematic variance in the DV as their most important data. That is, I will be calling into question the distinction between data and noise as these are understood in psychology and the cognitive sciences more broadly. To do this, I will need to describe how dynamical modeling works in the cognitive sciences and describe a set of experiments in my own laboratory as a case study in using noise as data. Along the way, I will grind various axes concerning dynamical models in the cognitive sciences as genuine explanations as opposed to "mere descriptions," and concerning the role of dynamical models in the context of discovery.

DYNAMICAL MODELING IN THE COGNITIVE SCIENCES

The research in the cognitive sciences that calls the methodological truism into question is a version of what is often called the *dynamical systems approach* to cognitive science (Kelso 1995; van Gelder 1995, 1998; Port and van Gelder 1996; Chemero 2009). In the dynamical systems approach, cognitive systems are taken to be dynamical systems, whose behavior is modeled using the branch of calculus known as dynamical systems theory. Dynamical modeling of the kind to be discussed later was introduced to psychology by Kugler, Kelso, and Turvey in 1980 as an attempt to make sense of Gibson's claim in 1979 that human action was regular but not regulated by any homunculus. They suggested that human limbs in coordinated action could be understood as nonlinearly coupled oscillators whose coupling requires energy to maintain and so tends to dissipate after a time. This suggestion is the beginning of coordination dynamics, the most well-developed form of the dynamical systems modeling in cognitive science.

The oft-cited locus classicus for dynamical research is work from the 1930s by Erich von Holst (1935/1973; see Kugler and Turvey 1987, and Iverson and Thelen 1999 for good accounts). Von Holst studied the coordinated movements of the fins of swimming fish. He modeled the fins as coupled oscillators, which had been studied by physicists since the seventeenth century when Huygens described the "odd sympathy" of pendulum clocks against a common wall. Von Holst assumed that on its own a fin has a preferred frequency of oscillation. When coupled to another fin via the fish's body and nervous systems, each fin tends to pull the other fin toward its preferred period of oscillation and will also tend to behave so as to maintain its own preferred period of oscillation. These are called the *magnet effect* and the *maintenance effect,* respectively. Von Holst observed that because of these interacting effects, pairs of fins on a single fish tended to reach a cooperative frequency of oscillation but with a high degree of variability around that frequency. One could model von Holst's observations with the following coupled equations:

$$\dot{\theta}_{fin1} = \omega_{fin1} + \text{Magnet}_{2 \to 1}(\theta_{fin2} - \theta_{fin1})$$

$$\dot{\theta}_{fin2} = \omega_{fin2} + \text{Magnet}_{1 \to 2}(\theta_{fin1} - \theta_{fin2})$$

where ω is the preferred frequency of each fin and θ is the current frequency of each fin. In English, these equations say that the change in frequency of oscillation of each fin is a function of that fin's preferred frequency and the magnet effect exerted on it by the other fin.

There are three things to note about these equations. First, the first and second terms in the equations represent the maintenance and magnet effects, respectively. Second, the equations are nonlinear and coupled. You cannot solve the first equation without solving the second one simultaneously, and vice versa. Third, with these equations you can model the behavior of the two-fin system with just one parameter: the strength of the coupling between the fins (i.e., the strength of the magnet effect).

This basic model is the foundation of later coordination dynamics, especially in its elaboration by Haken, Kelso, and Bunz (1985; see also Kugler, Kelso, and Turvey 1980). The Haken-Kelso-Bunz (HKB) model elaborates the von Holst model by including concepts from synergetics as a way to explain self-organizing patterns in human bimanual coordination. In earlier experiments, Kelso (1984) showed that human participants who were asked to wag the index fingers on both of their hands along with a metronone invariably wagged them in one of two patterns. Either they wagged the fingers in phase, with the same muscle groups in each hand performing the same movement at the same time, or out of phase, with the same muscle groups in each hand "taking turns." Kelso found that out-of-phase movements were only stable at slower metronone frequencies; at higher rates, participants could not maintain out-of-phase coordination in their finger movements and transferred into in-phase movements. Haken, Kelso, and Bunz modeled this behavior using the following equation, a potential function

$$V(\phi) = -A\cos\phi - 2B\cos 2\phi$$

where ϕ is the relative phase of the oscillating fingers and $V(\varphi)$ is the energy required to maintain relative phase ϕ. In-phase finger wagging has a relative phase of 0; out-of-phase finger wagging is relative phase π. The relative phase is a collective variable capturing the relationship between the motions of the two fingers. The transition from relative phase π to relative phase 0 at higher metronome frequencies is modeled by this derivative of relative phase:

$$d\phi / dt = -A\sin\phi - 2B\sin 2\phi$$

In these equations the ratio B/A is a control parameter, a parameter for which quantitative changes to its value lead to qualitative changes in the overall system. These equations describe all the behavior of the system.

The behavior of the coordinated multilimb system can be seen from Figure 2.1, which shows the system potential at various values of the control parameter. At the top of the figure, when the metronome frequency is low and the control parameter value is near 1, there are two attractors: a deep one at $\varphi = 0$ and relatively shallow one at $\varphi = \pm \pi$. As the value of the control parameter decreases (moving down the figure), the attractor at $\pm\pi$ gets shallower, until it disappears at 0.25. This is a critical point in the behavior system, a point at which qualitative behavior changes.

The HKB model accounts for the the behavior of this two-part coordinated system. Given the model's basis in synergetics, it makes a series of other predictions about finger wagging, especially concerning system behavior around critical points in the system. All these predictions have been verified in further experiments. Moreover, the HKB model does not apply

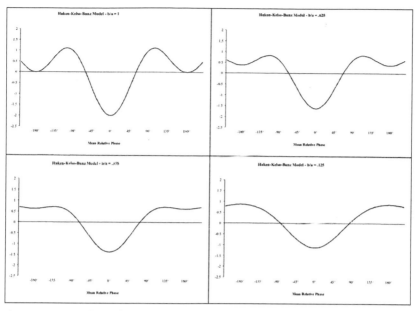

Figure 2.1. Attractor layout for the Haken-Kelso-Bunz model.

just to finger wagging. In the years since the original HKB study was published it has been extended many times and applied to a wide variety of phenomena in the cognitive sciences. That is, HKB has served as a *guide to discovery* (Chemero 2009) in that it has inspired a host of new testable hypotheses in the cognitive sciences. A brief survey of the contributions of the HKB model and its extensions to the cognitive and neurosciences would include the following:

- *Attention* (e.g., Amazeen et al. 1997; Temprado et al. 1999)
- *Intention* (e.g., Scholz and Kelso 1990)
- *Learning* (e.g., Amazeen 2002; Zanone and Kostrubiec 2004; Newell, Liu, and Mayer-Kress 2008)
- *Handedness* (e.g., Treffner and Turvey 1996)
- *Polyrhythms* (e.g., Buchanan and Kelso 1993; Peper, Beek, and van Wieringen 1995; Sternad, Turvey, and Saltzman 1999)
- *Interpersonal coordination* (Schmidt and Richardson 2008)
- *Cognitive modulation of coordination* (Pellecchia, Shockley, and Turvey 2005; Shockley and Turvey 2005)

- *Sentence processing* (e.g., Olmstead et al. 2009)
- *Speech production* (Port 2003)
- *Brain–body coordination* (Kelso et al. 1998)
- *Neural coordination dynamics* (e.g., Jirsa, Fuchs, and Kelso 1998; Bressler and Kelso 2001; Thompson and Varela 2001; Bressler 2002)

Coordination dynamics has been the most visible of the fields of application of dynamical models in the cognitive sciences, but it has hardly been the only one. Dynamical modeling has been applied to virtually every phenomenon studied in the cognitive sciences. It is well established and here to stay.

DYNAMICAL EXPLANATION IN THE COGNITIVE SCIENCES

Although not everyone would agree with me, I think that the biggest mistake I made in my 2009 book was to talk about dynamical modeling as generating explanations with no real argument that dynamical models in cognitive science could be explanations. Essentially, I underestimated the popularity of the "dynamical models are mere descriptions, not explanations" attitude toward dynamical modeling. Generally, this objection to dynamical models is made by claiming that dynamical models can only serve as covering law explanations, and rehearsing objections to covering law explanations. The reasons to object to counting dynamical models as explanations can stem from a strong theoretical commitment to computational explanation (e.g., Dietrich and Markman 2003; Adams and Aizawa 2008) or from a normative commitment to mechanistic philosophy of science (e.g., Craver 2007; Kaplan and Bechtel 2011; Kaplan and Craver 2011). Yet many cognitive scientists are committed to neither computational explanation nor normative mechanistic philosophy of science, and they can embrace dynamical explanations as genuine explanations. There is good reason to take dynamical explanations to be genuine explanations and not as mere descriptions.

Of course, dynamical explanations do not propose a causal mechanism that is shown to produce the phenomenon in question. Rather, they show that the change over time in the set of magnitudes in the world can be captured by a set of differential equations. These equations are law-like, and in some senses dynamical explanations are similar to *covering law explanations* (Bechtel 1998; Chemero 2000, 2009). That is, dynamical explanations show that particular phenomena could have been predicted, given local conditions and some law-like general principles.

In the case of HKB and coordination of limbs, we predict the behavior of the limbs, using the mathematical model and the settings of various parameters. Notice too that this explanation is counterfactual supporting. We can make predictions about which coordination patterns would be observed in settings that the system is not in fact in. The similarities between dynamical explanation and covering law explanation have led some commentators to argue that dynamical explanation is a kind of *predictivism,* according to which phenomena are explained if they could have been predicted. This claim that dynamical modeling is a kind of predictivism is followed by a two-part argument:

1. Rehearsals of objections to predictivism, and then
2. Claims that, because of these objections, dynamical models cannot be genuine explanations (Kaplan and Bechtel 2011; Kaplan and Craver 2011).

This might be an appropriate way to respond to claims that dynamical models were a genuine variety of explanations if they were written in reponse to works in which the similarities between dynamical modeling and covering law explanation were stressed (Bechtel 1998; Chemero 2000, 2009). However, the two-part argument has been made in response to discussions of dynamical models in which they were not characterized as covering law explanations (Chemero and Silberstein 2008; Stepp, Chemero, and Turvey 2011). In fact, Stepp and colleagues (2011) focus on the way that dynamical models provide explanatory *unification* in a sense similar to that discussed by Friedman (1974) and Kitcher (1989). (Richardson 2006 makes a similar point about biological phenomena.)

In the best dynamical explanations, an initial model of some phenomenon is reused in slightly altered form so that apparently divergent phenomena are brought under a small group of closely related models. This can be seen from the phenomena discussed earlier in "Dynamical Modeling in the Cognitive Sciences." The cognitive phenomena accounted for by the HKB model have been previously thought to be unrelated to one another; standard cognitive phenomena such as attention and problem solving are shown to be of a piece with social interaction, motor control, language use, and neural activity. All these are accounted for with the same basic model, and as we noted earlier this model was derived from synergetics. Given this, these cognitive phenomena are also unified with other phenomena to which synergetic models have been applied—self-organizing patterns of activity in systems far from thermodynamic equilibrium. So the applications of HKB in cognitive science not only bring these apparently disparate areas of cognitive science together, they also unify these parts of cognitive science with the areas of physics, chemistry, biology, ecology, sociology, and economics that have been modeled using synergetics (see Haken 2007).

Notice, however, that the unification achieved by dynamical models such as HKB differs from that described by Friedman and Kitcher. According to both Friedman and Kitcher, unification is a way of achieving scientific reduction; as Kitcher (1989) puts it, it is a matter of reducing the number of "ultimate facts." No such aim is served by the sort of unification one sees in dynamical modeling of the kind we have described. This is the case because the law-like generalizations of synergetics (like HKB) are *scale-invariant*. They are not characteristic of any particular scale and thus do not imply anything like traditional reductionism, whether ontological, intertheoretic, or in the store of ultimate facts. In the variety of unification achieved by dynamical models in cognitive science, the *only* thing that is reduced is the number of explanatory strategies employed. For this reason, Michael Turvey (1992) has called the strategy of consistently reapplying variants of HKB to new phenomena "strategic reductionism." More recently, Michael Silberstein and I have described this aspect of dynamical explanation in the cognitive sciences as *nonreductive unification* (Silberstein and Chemero 2013).

The point of this discussion is that dynamical models do a wide array of explanatory work. They do provide descriptions, and they do allow for predictions, but they also serve the epistemic purpose of unifying a wide range of apparently dissimilar phenomena both within the cognitive sciences and between the cognitive sciences and other sciences under a small number of closely related explanatory strategies. They also serve as guides to discovery. Arguably, dynamical models do all the explanatory work that scientists are after with the exception of describing mechanisms. Only a commitment to a definitional relationship between explanation and mechanism should prevent one from accepting that the dynamical models in the cognitive sciences are genuine explanations.

This claim that dynamical explanations in cognitive science are genuine explanations is in no way intended to disparage reductionistic or mechanistic explanations. Cognitive science will need dynamical, mechanistic, and reductionistic explanations (Chemero and Silberstein 2008; Chemero 2009). To borrow a metaphor from Rick Dale, we would be puzzled if someone told us that they had an explanation of the Mississippi River (Dale 2008; Dale,

Dietrich, and Chemero 2009). Because the Mississippi River is such a complex entity, we would want to know what aspect of the river they are explaining. The sedimentation patterns? The economic impact? The representation in American literature? Surely these disparate aspects of the Mississippi could not have just one explanation or even just one kind of explanation. Cognition, I submit, is just as complicated as the Mississippi River, and it would be shocking if just one style of explanation could account for all of it. For this reason, it is wise to adopt explanatory pluralism in the cognitive sciences.

BRIEF INTERLUDE ON PHILOSOPHICAL ISSUES IN MODELING

It is worth noting here that dynamical modeling and dynamical explanation in the cognitive sciences, as just described, occupy a rather different place than modeling and simulation in other sciences. In the cognitive sciences, dynamical models are not parts of the theory, developments of the theory, or abstract experiments. Kelso, Kugler, and Turvey wanted to develop an understanding of action that was consonant with Gibson's ecological psychology and Gibson's claim that action is regular without being regulated. Nothing in these basic aims or in ecological psychology required modeling action as self-organization in systems far from equilibrium, which is why Kelso, Kugler, and Turvey's decision to use these models is such an important insight. These models are not part of ecological psychology, and they were not derived from ecological psychology. Indeed, they were imported from another scientific discipline. Moreover, these models brought a host of assumptions and techniques along with them. It was these assumptions and techniques that became the engines for designing experiments and refining the models as they have been applied in the cognitive sciences.

The HKB model and later modifications of it are, as Morrison (2007) and Winsberg (2009) would say, autonomous. In fact, they are autonomous in a way that is stronger than what Morrison and Winsberg have in mind. The form of synergetics-based mathematical models that Kelso, Kugler, and Turvey import from physics are not only not developments of ecological psychological theory, they have a kind of independence from *any* theory, which is why they could move from physics to the cognitive sciences. This is a disagreement with Winsberg (2009), who claims that models do not go beyond the theories in which they are put to use. It is this independence that allows the models to serve as unifying explanations of phenomena, as discussed earlier in "Dynamical Explanation in the Cognitive Sciences." This is a disagreement with Morrison (2000), who argues that explanation and unification are separate issues.

Perhaps part of the reason for these differences with Morrison and Winsberg stems from the fact that they both focus on areas of research within the physical sciences, in which theories often just take the form of nonlinear differential equations. When this is the case, one of the key roles for models is to render the equations of the theory into solvable and computationally tractable form. This is dramatically different from the role models play in the cognitive sciences, whose theories are rarely mathematical in nature. Gibson's articulation of ecological psychology in his posthumous book (1979), which inspired Kelso, Kugler, and Turvey, had not a single equation.

It is important to note here that the claims I make in this section, indeed in this whole essay, apply only to dynamical modeling of a particular kind in a particular scientific enterprise. Models play different roles in different sciences and even in different parts of individual sciences. Philosophers of science should be modest, realizing that their claims about the relationship of models to theory and experiment might not hold in sciences other than the one from which they are drawing their current example. The relationship might not hold in a different subfield of the science they are currently discussing, and maybe not in a different research team in the same subfield. Just as it is important for cognitive scientists to acknowledge explanatory pluralism, it is important to acknowledge a kind of philosophy of science pluralism: different sciences do things differently, and lots of claims about capital-s-Science will fail to apply across the board.

COMPLEX SYSTEMS PRIMER

As noted previously, the HKB model makes a series of predictions about the coordinated entities to which it is applied. One of these predictions is that there will be *critical fluctuations* as the coordinated system approaches a critical point—for example, just before it transitions from out-of-phase to in-phase coordination patterns. This prediction of critical fluctuation, which has been verified in nearly all the areas of human behavior and cognition listed in the dynamical modeling discussion, is perhaps the clearest way in which the use of HKB allies the cognitive sciences with the sciences of complexity. In particular, it suggests that coordinated human action arises in an *interaction-dominant system* (van Orden, Holden, and Turvey 2003, 2005; Riley and Holden 2012). It has also given rise to new research programs in the cognitive sciences and neurosciences concerning the character of fluc-tuations, a research program that promises to shed light on questions about cognition and the mind that previously seemed "merely philosophical." In the remainder of this section I will briefly (and inadequately) set out the basic concepts and methodologies of this new research program. In the next section, I will show how, in a particular instance, those concepts and methodologies have been put into use.

First, some background: an *interaction-dominant system* is a highly inter-connected system, each of whose components alters the dynamics of many of the others to such an extent that the effects of the interactions are more powerful than the intrinsic dynamics of the components (van Orden, Holden, and Turvey 2003). In an interaction-dominant system, inherent variability (i.e., fluctuations or noise) of any individual component C propa-gates through the system as a whole, altering the dynamics of the other components, say, D, E, and F. Because of the dense connections among the components of the systems, the alterations of the dynamics of D, E, and F will lead to alterations to the dynamics of component C. That initial ran-dom fluctuation of component C, in other words, will reverberate through the system for some time. So too, of course, would nonrandom changes in the dynamics of component C. This tendency for reverberations gives interaction-dominant systems what is referred to as "long memory" (Chen, Ding, and Kelso 1997).

In contrast, a system without this dense dynamical feedback, a *component-dominant system*, would not show this long memory. For example, imagine a computer program that controls a robotic arm. Although noise that creeps into the commands sent from the computer to the arm might lead to a weld that misses its intended mark by a few millimeters, that missed weld will not alter the behavior of the program when it is time for the next weld. Component-dominant systems do not have long memory. Moreover, in interaction-dominant systems one cannot treat the components of the system in isolation: because of the widespread feedback in interaction-dominant systems, one cannot isolate components to determine exactly what their contribution is to particular behavior. And because the effects of interactions are more powerful than the intrinsic dynamics of the compo-nents, the behavior of the components in any particular interaction-dominant system is not predictable from their behavior in isolation or from their be-havior in some other interaction-dominant system. Interaction-dominant systems, in other words, are not *modular*. Indeed, they are in a deep way *unified* in that the responsibility for system behavior is distributed across all the components. As will be discussed later, this unity of interaction-dominant systems has been used to make strong claims about the nature of human cognition (van Orden, Holden, and Turvey 2003, 2005; Holden, van Orden, and Turvey 2009; Riley and Holden 2012).

In order to use the unity of interaction-dominant systems to make strong claims about cognition, it has to be argued that cognitive systems are, at least sometimes, interaction dominant. This is where the critical fluctuations posited by the HKB model come in. Interaction-dominant systems self-organize, and some will exhibit *self-organized criticality*. Self-organized criticality is the tendency of an interaction-dominant system to maintain itself near critical boundaries so that small changes in the system or its environment can lead to large changes in overall system behavior. It is easy to see why self-organized criticality would be a useful feature for behavioral control: a system near a critical boundary has built-in behavioral flexibility because it can easily cross the boundary to switch behaviors (e.g., going from out-of-phase to in-phase coordination patterns). (See Holden, van Orden, and Turvey 2009 and Kello, Anderson, Holden, and van Orden 2008 for further discussion.)

It has long been known that self-organized critical systems exhibit a special variety of fluctuation called *1/f noise* or *pink noise* (Bak, Tang, and Wieseneld 1988). That is, 1/f noise or pink noise is a kind of not-quite-random, correlated noise, halfway between genuine randomness (white noise) and a drunkard's walk, in which each behavior is constrained by the prior one (brown noise). Often 1/f noise is described as a fractal structure in a time series, in which the variability at a short time scale is correlated with variability at a longer time scale. In fact, interaction-dominant dynamics predicts that this 1/f noise would be present. As discussed previously, the fluctuations in an interaction-dominant system percolate through the system over time, leading to the kind of correlated structure to variability that is 1/f noise. Here, then, is the way to infer that cognitive and neural systems are interaction dominant: exhibiting 1/f noise is evidence that a system is interaction dominant.

This suggests that the mounting evidence that 1/f noise is ubiquitous in human physiological systems, behavior, and neural activity is also evidence that human physiological, cognitive, and neural systems are interaction dominant. To be clear, there are ways other than interaction-dominant dynamics to generate 1/f noise. Simulations show that carefully gerrymandered component-dominant systems can exhibit 1/f noise (Wagenmakers, Farrell, and Ratcliff 2005). But such gerrymandered systems are not developed from any physiological, cognitive, or neurological principle and so are not taken to be plausible mechanisms for the widespread $1/f$ noise in human physiology, brains, and behavior (van Orden, Holden, and Turvey 2005). So the inference from $1/f$ noise to interaction dominance is not foolproof, but there currently is no plausible explanation of the prevalence of $1/f$ noise other than interaction dominance. Better than an inference to the best explanation, that $1/f$ noise indicates interaction-dominant dynamics is an inference to the only explanation.

HEIDEGGER IN THE LABORATORY

Dobromir Dotov, Lin Nie, and I have recently used the relationship between $1/f$ noise and interaction dominance to empirically test a claim that can be derived from Heidegger's phenomenological philosophy (1927/1962; see Dotov, Nie, and Chemero 2010; Nie, Dotov, and Chemero 2011). In particular, we wanted to see whether we could gather evidence supporting the transition Heidegger proposed for "ready-to-hand" and "unready-to-hand" interactions with tools. Heidegger argued that most of our experience of tools is unreflective, smooth coping with them. When we ride a bicycle competently, for example, we are not aware of the bicycle but of the street, the traffic conditions, and our route home. The bicycle itself recedes in our experience and becomes the thing through which we experience the road. In Heidegger's language, the bicycle is *ready-to-hand,* and we experience it as a part of us, no different than our feet pushing the pedals. Sometimes, however, the derailleur does not respond right away to an attempt to shift gears or the brakes grab more forcefully than usual, so the bicycle becomes temporarily prominent in our experience—we notice the bicycle. Heidegger would say that the bicycle has become *unready-to-hand,* in that our smooth use of it has been interrupted temporarily and it has become, for a short time, the object of our experience. (In a more permanent breakdown in which the bicycle becomes unuseable, like a flat tire, we experience it as present-at-hand.)

To gather empirical evidence for this transition, we relied on dynamical modeling that uses the presence of $1/f$ noise or pink noise as a signal of sound physiological functioning (e.g., West 2006). As noted previously, $1/f$ noise or pink noise is a kind of not-quite-random, correlated noise, halfway between genuine randomness (white noise) and a drunkard's walk, in which each fluctuation is constrained by the prior one (brown noise). Often $1/f$ noise is described as a fractal structure in a time series in which the variability at a short time scale is correlated with variability at a longer time scale. (See Riley and Holden 2012 for an overview.) The connection between sound physiological functioning and $1/f$ noise allows for a prediction related to Heidegger's transition: when interactions with a tool are ready-to-hand, the human-plus-tool should form a single interaction-dominant system; this human-plus-tool system should exhibit $1/f$ noise.

We showed this to be the case by having participants play a simple video game in which they used a mouse to control a cursor on a monitor. Their task was to move the cursor so as to "herd" moving objects to a circle on the center of the monitor. At some point during the trial, the connection between the mouse and cursor was temporarily disrupted so that movements of the cursor did not correspond to movements of the mouse, before returning to normal. The primary variable of interest in the experiment was the accelerations of the hand-on-the-mouse. Accelerations are of special interest because they correspond to purposeful actions on the part of the participant, when she changes the way her hand is moving. Instantaneous accelerations in three dimensions were measured at 875 Hz. We used only horizontal accelerations (i.e., left to right), using a 60 Hz low-pass filter to eliminate variability that was not from behavior, and then we down-sampled the data by a factor of 10. This gave us a time series of accelerations. This time series was subjected to *detrended fluctuation analysis* (DFA). In DFA, the time series is divided into temporal intervals of sizes varying from one second to the size of the whole time series. For each interval size, the time series is divided into intervals of that size, and within each of those intervals a trend line is calculated. The fluctuation function $F(n)$ for each interval size n is the root mean square of the variability from the trend line within each interval of that size. $F(n)$ is calculated for intervals of 1, 2, 4, 8, 16 . . . seconds. As you would expect, $F(n)$ increases as the interval size increases.

The values of $F(n)$ were plotted against the interval size in a log-log plot (Figure 2.2). The slope of the trend line in that log-log plot is the primary output of DFA, called a Hurst exponent. Hurst exponents of approximately 1 indicate the presence of $1/f$ noise, with lower values indicating white noise and higher values indicating brown noise. The raw data are folded, spindled, and mutilated to determine whether the system exhibits $1/f$ noise and can be assumed to be interaction dominant.

As predicted, the hand–mouse movements of the participants exhibited $1/f$ noise while the video game was working correctly. The $1/f$

noise decreased, almost to the point of exhibiting pure white noise, during the mouse perturbation. So while the participants were smoothly playing the video game, they were part of a human–computer interaction-dominant system; when we temporarily disrupted performance, that pattern of variability temporarily disappeared. This is evidence that the mouse was experienced as ready-to-hand while it was working correctly and became

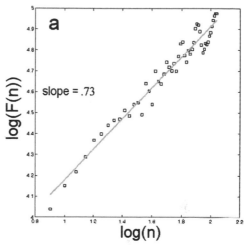

Figure 2.2. Detrended fluctuation analysis (DFA) plot for a representative participant in Dotov, Nie, and Chemero (2010).

unready-to-hand during the perturbation. (See Dotov, Nie, and Chemero 2010; Nie, Dotov, and Chemero 2011; and Dotov et al. 2017.)

It is important to note the role of the dynamical modeling. The dynamical model was used to put Heidegger's claims about phenomenology into touch with potentially gatherable data. In this case, after several unsuccessful attempts to demonstrate Heidegger's ideas and the thesis of extended cognition, we learned about the use of detrended fluctuation analysis and $1/f$ noise. Understanding the claims that could be made using detrended fluctuation analysis led us to design our experimental task in the way that we did. In effect, the models made Heidegger's claims empirically accessible and allowed us to gather evidence for them. Moreover, the results from prior dynamical modeling acted as a *guide to discovery*, a source of new hypotheses for further experimental testing (Chemero 2000, 2009). In this case, the findings concerning $1/f$ noise and physiology and the knowledge of the way $1/f$ noise could be detected inspired the experimental task. This makes clear that the dynamical models are not acting as mere descriptions of data nor even only as tools for analyzing data after they have been gathered; rather, dynamical models play a significant role in hypothesis generation. This is the point that is apparent especially to those who are both modelers and philosophers of science, such as Isabelle Peschard (2007).

DATA AND NOISE, AND NOISE

For current purposes, the findings about Heidegger's phenomenology and role of mathematical modeling in hypothesis generation are of secondary importance. What matters more is what happened to the data along the way. As noted earlier, the data we gathered concerned the accelerations of a hand on a computer mouse. Though these movements were gathered in three dimensions, only one (side-to-side) dimension was used. Of the other two dimensions, only one was significant: the mouse moved along the surface of a table, so up-down accelerations would be negligible. But one would think that the third dimension (front-back) would be very important.

All the participants were moving their hands in every direction along the surface of the table, and only very rarely were they moving them purely side-to-side or front-back. So the acceleration data, as we collected it, does not capture the actual behavior of the participants. We could, of course, have recovered the actual two-dimensional instantaneous accelerations computationally, but we did not. The reason we did not was because we did not care about how the participants moved their hands while they played the video game; we also did not care how well they played the video game. What we cared about were the fluctuations in their hand movements rather than the actual hand movements. That is, we cared about the noise, not the data.

Indeed, getting rid of the data and leaving only the noise is the purpose of detrended fluctuation analysis. It is a step-by-step procedure to eliminate anything about the time series other than the noise, and then to determine the character of that noise, whether it is white, pink, or brown. Consider again the detrending part of DFA: within each time interval, the linear trend is determined, and that trend line is subtracted from each instantaneous acceleration. If we cared about the actual hand movements, the trend line itself would be the object of interest because it would let us know how a participant moved her hand on the mouse during any particular time interval. But the actual movements of the hand on the mouse are not relevant to answering the questions we wanted to answer, that the methods of complex systems science allowed us to answer. That is, by ignoring the data and focusing on the noise we were able to argue that the participant–computer system formed a single interaction-dominant system, and to provide some empirical confirmation of a portion of Heidegger's phenomenology.

Given the rate at which methods from complex systems methods are being brought into cognitive sciences, we have good reason to doubt the general truth of the methodological truism that psychology students are commonly taught—that is, that a good experiment maximizes primary variance and minimizes noise. This is certainly an issue that psychologists and psychology teachers ought to be concerned with. Of course, at an introductory level, many complexities are papered over, often so much so that what we tell students is not strictly speaking true. Unfortunately, many practicing scientists take what I call the methodological truism to be a truth. It is not. Really, the widespread use of complex systems methods should introduce large changes to the way experiments are designed in the cognitive sciences. This is the case because analyses such as DFA involve a violation of the assumptions of the generalized linear model that undergirds the statistics used to analyze their outputs (such as analysis of variance, multivariate analysis of variance, regressions, etc.). In particular, because DFA is applied to systems with "long memory" (i.e., systems in which the outcome of any particular observation is connected to the outcome of other observations) the observations are not independent of one another as is required by the generalized linear model. As just noted, this is an issue well outside the scope of what an introductory-level student needs to know, but it is an issue nonetheless.

What is more of an issue for philosophers of science is that we have good reason to call the distinction between data and noise into question. This is problematic because that distinction is so deeply built into the practice of cognitive scientists—and indeed all scientists. There are error bars on most graphs and noise terms in most mathematical models. In every data set, that is, the assumption is that there is something like what Bogen and Woodward (1988) call the phenomenon—which the scientist cares about—and the noise—which she does not care about. When complex systems methods are used in the cognitive sciences, as we have seen, noise is the phenomenon.

Bogen (2010) makes a similar point that sometimes the noise is what scientists care about. This much is clear from Figure 2.2. Recall that DFA produces a log-log plot of fluctuations against interval sizes. The slope of a trend line on that plot is the Hurst exponent, the value of which lets you know whether $1/f$ noise is present. Very few of the points shown in Figure 2.2 are actually on that trend line, and we could easily calculate the goodness of fit of that trend line. That is, after we have removed all the data about the actual hand movements using DFA and made the character of the noise in hand movements the object of our interest, there is still further noise. So when the phenomenon the scientist cares about is noise, there is second-order noise that she does not care about.

It seems that one could re-establish the distinction between data and noise, but only in terms of the explanatory interests of scientists. Data are the outcomes of observations, measurements, and manipulations that the scientist cares about. Sometimes these outcomes will be measures of fluctuations, what has traditionally been called noise. Noise, then, will be variability in the outcomes of those measurements and manipulations that the scientist does not care about. If a scientist (or someone like me who occasionally pretends to be a scientist) cares about confirming some claims derived from Heidegger's phenomenology by showing that humans and computers together can comprise unified, interaction-dominant systems, he will take Hurst exponents to be data, even though they are measurements of noise. He will then take error in the linear regression used to produce the Hurst exponent as noise. Understanding noise this way—as ignored variability in what the scientist cares about—makes distinguishing data from noise of a piece with the problem of *relevance* that Peschard discusses (2010, 2012). Whether something is data or noise depends on what scientists are interested in, and what they think matters to the phenomena they are exploring.

THE METHODOLOGICAL TRUISM, AGAIN

The cognitive sciences are young sciences, with a creation mythology that places their beginning in 1956. Psychology is a bit older, having begun in earnest in the late nineteenth century. Yet it seems that cognitive scientists and psychologists have already gathered the majority of the low-hanging fruit, and the interesting findings that can be had with timing experiments and *t*-tests have mostly been found. It is for this reason that new experimental and analytical techniques are being brought into the cognitive sciences from the sciences of complexity.

In this essay, I have looked at some consequences of the import into the cognitive sciences of tools from synergetics. There are, I have suggested, consequences both for the cognitive sciences and for the philosophy thereof. I want to close by reconsidering what I called the "methodological truism" in the first section. I repeat it here.

The goal of a good experiment is to maximize primary variance, eliminate secondary variance, and minimize error variance.

What I hope to have shown in this essay is that teaching students in introductory psychology the methodological truism is a lot like telling young children that babies are delivered by storks. It is something that was thought to be good enough until the hearers were mature enough to know the messy, complicated truth about things. Unfortunately, the story is not just untrue, but it covers up all the interesting stuff. And, of course, we adults should know better.

REFERENCES

Adams, Fredrick, and Kenneth Aizawa. 2008. *The Bounds of Cognition*. Malden, Mass.: Blackwell.

Amazeen, Eric L., Polemnia G. Amazeen, Paul J. Treffner, and Michael T. Turvey. 1997. "Attention and Handedness in Bimanual Coordination Dynamics." *Journal of Experimental Psychology: Human Perception and Performance* 23: 1552–60.

Amazeen, Polemnia G. 2002. "Is Dynamics the Content of a Generalized Motor Program for Rhythmic Interlimb Coordination?" *Journal of Motor Behavior* 34: 233–51.

Bak, Per, Chao Tang, and Kurt Wiesenfeld. 1988. "Self-Organized Criticality." *Physical Review A* 38: 364–74.

Bechtel, William. 1998. "Representations and Cognitive Explanations: Assessing the Dynamicist Challenge in Cognitive Science." *Cognitive Science* 22: 295–318.

Bogen, James. 2010. "Noise in the World." *Philosophy of Science* 77: 778–91.

Bogen, James, and James Woodward. 1988. "Saving the Phenomena." *Philosophical Review* 97: 303–52.

Bressler, Steven L. 2002. "Understanding Cognition through Large-Scale Cortical Networks." *Current Directions in Psychological Science* 11: 58–61.

Bressler, Steven L., and J. A. Scott Kelso. 2001. "Cortical Coordination Dynamics and Cognition." *Trends in Cognitive Sciences* 5: 26–36.

Buchanan, J. J., and J. A. Scott Kelso. 1993. "Posturally Induced Transitions in Rhythmic Multijoint Limb Movements." *Experimental Brain Research* 94: 131–42.

Chemero, Anthony. 2000. "Anti-representationalism and the Dynamical Stance." *Philosophy of Science* 67: 625–47.

Chemero, Anthony. 2009. *Radical Embodied Cognitive Science*. Cambridge, Mass.: MIT Press.

Chemero, Anthony, and Michael Silberstein. 2008. "After the Philosophy of Mind: Replacing Scholasticism with Science." *Philosophy of Science* 75: 1–27.

Chen, Yanqing, Mingzhou Ding, and J. A. Scott Kelso. 1997. "Long Memory Processes $1/f^{\alpha}$ Type in Human Coordination." *Physical Review Letters* 79: 4501–4.

Craver, Carl. 2007. *Explaining the Brain: What a Science of the Mind-Brain Could Be.* New York: Oxford University Press.

Dale, Rick. 2008. "The Possibility of a Pluralist Cognitive Science." *Journal of Experimental and Theoretical Artificial Intelligence* 20: 155–79.

Dale, Rick, Eric Dietrich, and Anthony Chemero. 2009. "Pluralism in the Cognitive Sciences." *Cognitive Science* 33: 739–42.

Dietrich, Eric, and Arthur B. Markman. 2003. "Discrete Thoughts: Why Cognition Must Use Discrete Representations." *Mind and Language* 18: 95–119.

Dotov, Dobromir G., Lin Nie, and Anthony Chemero. 2010. "A Demonstration of the Transition from Ready-to-Hand to Unready-to-Hand." *PLoS ONE* 5: e9433.

Dotov, Dobromir G., Lin Nie, Kevin Wojcik, Anastasia Jinks, Xiaoyu Yu, and Anthony Chemero. 2017. "Cognitive and Movement Measures Reflect the Transition to Presence-at-Hand." *New Ideas in Psychology* 45: 1–10.

Franklin and Marshall College Psychology Department. *Introduction to Psychology Laboratory Manual.* Unpublished.

Friedman, Michael. 1974. "Explanation and Scientific Understanding." *Journal of Philosophy* 71: 5–19.

Gibson, James J. 1979. *The Ecological Approach to Visual Perception.* Boston: Houghton-Mifflin.

Haken, Hermann. 2007. "Synergetics." *Scholarpedia* 2: 1400.

Haken, Hermann, J. A. Scott Kelso, and H. Bunz. 1985. "A Theoretical Model of Phase Transitions in Human Hand Movements." *Biological Cybernetics* 51: 347–56.

Heidegger, Martin. (1927) 1962. *Being and Time.* Translated by J. Macquarrie and E. Robinson. New York: Harper & Row. First published in 1927 as *Sein und Zeit.*

Holden, John G., Guy C. Van Orden, and Michael T. Turvey. 2009. "Dispersion of Response Times Reveals Cognitive Dynamics." *Psychological Review* 116: 318–42.

Iverson, Jana M., and Esther Thelen. 1999. "Hand, Mouth, and Brain: The Dynamic Emergence of Speech and Gesture." *Journal of Consciousness Studies* 6: 19–40.

Jirsa, Viktor K., Armin Fuchs, and J. A. Scott Kelso. 1998. "Connecting Cortical and Behavioral Dynamics: Bimanual Coordination." *Neural Computation* 10: 2019–45.

Kaplan, David M., and William Bechtel. 2011. "Dynamical Models: An Alternative or Complement to Mechanistic Explanations?" *Topics in Cognitive Science* 3: 438–44.

Kaplan, David M., and Carl Craver. 2011. "The Explanatory Force of Dynamical and Mathematical Models in Neuroscience: A Mechanistic Perspective." *Philosophy of Science* 78: 601–28.

Kello, Christopher T., Gregory G. Anderson, John G. Holden, and Guy C. van Orden. 2008. "The Pervasiveness of 1/*f* Scaling in Speech Reflects the Metastable Basis of Cognition." *Cognitive Science* 32: 1217–31.

Kelso, J. A. Scott. 1984. "Phase Transitions and Critical Behavior in Human Bimanual Coordination." *American Journal of Physiology—Regulatory, Integrative and Comparative Physiology* 246: R1000–4.

Kelso, J. A. Scott. 1995. *Dynamic Patterns: The Self-Organization of Brain and Behavior.* Cambridge, Mass.: MIT Press.

Kelso, J. A. Scott, A. Fuchs, T. Holroyd, R. Lancaster, D. Cheyne, and H. Weinberg. 1998. "Dynamic Cortical Activity in the Human Brain Reveals Motor Equivalence." *Nature* 23: 814–18.

Kitcher, Philip. 1989. "Explanatory Unification and the Causal Structure of the World." In *Scientific Explanation,* edited by Philip Kitcher and Wesley C. Salmon, 410–505. Minneapolis: University of Minnesota Press.

Kugler, Peter N., and Michael T. Turvey. 1987. *Information, Natural Law, and the Self-Assembly of Rhythmic Movement.* Hillsdale, N.J.: Erlbaum.

Kugler, Peter N., J. A. Scott Kelso, and Michael T. Turvey. 1980. "On the Concept of Coordinative Structures as Dissipative Structures. I. Theoretical Lines of Convergence." In *Tutorials in Motor Behavior,* edited by G. E. Stelmach and J. Requin, 3–47. Amsterdam: North Holland.

Morrison, Margaret. 2000. *Unifying Scientific Theories.* Cambridge: Cambridge University Press.

Morrison, Margaret. 2007. "Where Have All the Theories Gone?" *Philosophy of Science* 74: 195–228.

Newell, Karl M., Yeou-Teh Liu, and Gottfried Mayer-Kress. 2008. "Landscapes beyond the HKB Model." In *Coordination: Neural, Behavioral and Social Dynamics,* edited by Armin Fuchs and Viktor K. Jirsa, 27–44. Heidelberg: Springer-Verlag.

Nie, Lin, Dobromir G. Dotov, and Anthony Chemero. 2011. "Readiness-to-Hand, Extended Cognition, and Multifractality." In *Proceedings of the 33rd Annual Meeting of the Cognitive Science Society,* edited by Laura Carlson, Christoph Hölscher, and Thomas F. Shipley, 1835–40. Austin, Tex.: Cognitive Science Society.

Olmstead, Anne J., Navin Viswanathan, Karen A. Aicher, and Carol A. Fowler. 2009. "Sentence Comprehension Affects the Dynamics of Bimanual

Coordination: Implications for Embodied Cognition." *Quarterly Journal of Experimental Psychology* 62: 2409–17.

Pellecchia, Geraldine L., Kevin Shockley, and Michael T. Turvey. 2005. "Concurrent Cognitive Task Modulates Coordination Dynamics." *Cognitive Science* 29: 531–57.

Peper, C. Lieke E., Peter J. Beek, and Piet C. W. van Wieringen. 1995. "Coupling Strength in Tapping a 2:3 Polyrhythm." *Human Movement Science* 14: 217–45.

Peschard, Isabelle. 2007. "The Values of a Story: Theories, Models, and Cognitive Values." *Principia* 11: 151–69.

Peschard, Isabelle. 2010. "Target Systems, Phenomena and the Problem of Relevance." *Modern Schoolman* 87: 267–84.

Peschard, Isabelle. 2012. "Forging Model/World Relations: Relevance and Reliability." *Philosophy of Science* 79: 749–60.

Port, Robert F. 2003. "Meter and Speech." *Journal of Phonetics* 31: 599–611.

Port, Robert F., and Timothy van Gelder. 1996. *Mind as Motion.* Cambridge, Mass.: MIT Press.

Richardson, Robert C. 2006. "Explanation and Causality in Self-Organizing Systems." In *Self-Organization and Emergence in Life Sciences,* edited by Bernard Feltz, Marc Crommelinck, and Philippe Goujon, 315–40. Dordrecht, the Netherlands: Springer.

Riley, Michael A., and John G. Holden. 2012. "Dynamics of Cognition." *WIREs Cognitive Science* 3: 593–606.

Schmidt, Richard C., and Michael J. Richardson. 2008. "Dynamics of Interpersonal Coordination." In *Coordination: Neural, Behavioral and Social Dynamics,* edited by Armin Fuchs and Viktor K. Jirsa, 281–308. Heidelberg: Springer-Verlag.

Scholz, J. P., and J. A. Scott Kelso. 1990. "Intentional Switching between Patterns of Bimanual Coordination Depends on the Intrinsic Dynamics of the Patterns." *Journal of Motor Behavior* 22: 98–124.

Shockley, Kevin, and Michael T. Turvey. 2005. "Encoding and Retrieval during Bimanual Rhythmic Coordination." *Journal of Experimental Psychology: Learning, Memory, and Cognition* 31: 980–90.

Silberstein, Michael, and Anthony Chemero. 2013. "Constraints on Localization and Decomposition as Explanatory Strategies in the Biological Sciences." *Philosophy of Science* 80: 958–70.

Stepp, Nigel, Anthony Chemero, and Michael T. Turvey. 2011. "Philosophy for the Rest of Cognitive Science." *Topics in Cognitive Science* 3: 425–37.

Sternad, Dagmar, Michael T. Turvey, and Elliot L. Saltzman. 1999. "Dynamics of 1:2 Coordination: Generalizing Relative Phase to n:m Rhythms." *Journal of Motor Behavior* 31: 207–24.

Temprado, Jean-Jacques, Pier-Giorgio Zanone, Audrey Monno, and Michel Laurent. 1999. "Attentional Load Associated with Performing and Stabilizing Preferred Bimanual Patterns." *Journal of Experimental Psychology: Human Perception and Performance* 25: 1579–94.

Thompson, Evan, and Francisco J. Varela. 2001. "Radical Embodiment: Neural Dynamics and Consciousness." *Trends in Cognitive Sciences* 5: 418–25.

Treffner, P. J., and M. T. Turvey. 1996. "Symmetry, Broken Symmetry, and Handedness in Bimanual Coordination Dynamics." *Experimental Brain Research* 107: 163–78.

Turvey, Michael T. 1992. "Affordances and Prospective Control: An Outline of the Ontology." *Ecological Psychology* 4: 173–87.

van Gelder, Timothy. 1995. "What Might Cognition Be if Not Computation?" *Journal of Philosophy* 91: 345–81.

van Gelder, Timothy. 1998. "The Dynamical Hypothesis in Cognitive Science." *Behavioral and Brain Sciences* 21: 615–28.

Van Orden, Guy C., John G. Holden, and Michael T. Turvey. 2003. "Self-Organization of Cognitive Performance." *Journal of Experimental Psychology: General* 132: 331–51.

Van Orden, Guy C., John G. Holden, and Michael T. Turvey. 2005. "Human Cognition and 1/f Scaling." *Journal of Experimental Psychology: General* 134: 117–23.

Von Holst, Erich. (1935) 1973. *The Behavioral Physiology of Animal and Man: The Collected Papers of Erich Von Holst,* vol. 1. Coral Gables, Fla.: University of Miami Press.

Wagenmakers, Eric-Jan, Simon Farrell, and Roger Ratcliff. 2005. "Human Cognition and a Pile of Sand: A Discussion on Serial Correlations and Self-Organized Criticality." *Journal of Experimental Psychology: General* 134: 108–16.

West, Bruce J. 2006. "Fractal Physiology, Complexity, and the Fractional Calculus." In *Fractals, Diffusion and Relaxation in Disordered Complex Systems,* edited by Y. Kalmykov, W. Coffey, and S. Rice, 1–92. Singapore: Wiley-Interscience.

Winsberg, Eric. 2009. *Science in the Age of Computer Simulation.* Chicago: University of Chicago Press.

Zanone, Pier-Giorgio, and Viviane Kostrubiec. 2004. "Searching for (Dynamic) Principles of Learning." In *Coordination Dynamics: Issues and Trends,* edited by Viktor K. Jiras and J. A. Scott Kelso, 57–89. Heidelberg: Springer Verlag.

3

ARTICULATING THE WORLD
Experimental Practice and Conceptual Understanding

JOSEPH ROUSE

QUESTIONS ABOUT THE INTERPRETATION of data are grounded in philosophical issues about the conceptual articulation of perceptual experience or causal interaction. I approach these issues by considering how experimental practice contributes to conceptual understanding. "Experimental" is used broadly to incorporate a wide variety of empirically oriented practices; observational sciences, clinical sciences, or comparative sciences such as paleontology or systematics are also "experimental" in this broad sense. "Experimental" as the umbrella term highlights that even seemingly descriptive or observational sciences typically must undertake material work (with instruments, sample collection and preparation, shielding from extraneous interference, observational protocols, and much more) to allow objects to show themselves appropriately. All empirical sciences intervene in and transform aspects of the world to let them be intelligible and knowable.

Despite renewed philosophical interest in experiment and material practice, conceptual articulation in science is still primarily understood as theory construction. Quine's (1953) famous image of scientific theory as a self-enclosed fabric or field that only encounters experience at its periphery is instructive. Quine not only neglects experimental activity in favor of perceptual receptivity ("surface irritations"; Quine 1960), but his image of conceptual development in the sciences involves a clear division of labor. Experience and experiment impinge from "outside" our theory or conceptual scheme to provide *occasions* for conceptual development. The resulting *work* of conceptual articulation is nevertheless a linguistic or mathematical activity of developing and regulating "internal" inferential relations among sentences or equations to reconstruct the "fabric" of theory.

This insistence upon the constitutive role of theory and theoretical language within scientific understanding arose from familiar criticisms of empiricist accounts of conceptual content. Many earlier empiricists in the philosophy of science are now widely recognized as making linguistic frameworks constitutive for how empirical data could have conceptual significance (Richardson 1998; Friedman 1999). Yet their "post-positivist" successors have gone further in this direction, and not merely by asserting that observation is theory laden. Most recent philosophical discussions of how theories or theoretical models relate to the world begin where phenomena have already been articulated conceptually. As one influential example, James Bogen and James Woodward (1988) argued,

> Well-developed scientific theories predict and explain facts about phenomena. Phenomena are detected through the use of data, but in most cases are not observable in any interesting sense of the term . . . Examples of phenomena, for which the above data might provide evidence, include weak neutral currents, the decay of the proton, and chunking and recency effects in human memory. (306)

Nancy Cartwright (1983, essay 7) recognizes that one cannot apply theories or models to events in the world without preparing a description of them in the proper terms for the theory to apply, but she only characterizes this "first stage of theory entry" as an operation on the "unprepared description [which] may well use the language and the concepts of the theory, but is not constrained by any of the mathematical needs of the theory" (133). Michael Friedman (1974) made this tendency to take conceptual articulation for granted especially clear by arguing that the "phenomena" scientific theories seek to explain are best understood as laws that characterize regular patterns rather than specific events. If the relation between theory and the world begins with the explanation of law-like patterns, however, that relation comes into philosophical purview only after the world has already been conceptualized.

Outside philosophy of science, how conceptual understanding is accountable to the world has gained renewed prominence from McDowell (1994), Brandom (1994), Haugeland (1998), and others. McDowell expresses shared concerns with the image of a treacherous philosophical passage. On one side looms the rocks of Scylla, where attempts to ground conceptual content on merely "Given" causal or experiential impacts run aground. On the other beckons the whirlpool Charybdis, where the entirely intralinguistic coherence of purported conceptual judgments would become a mere "frictionless spinning in a void."[1] The post-positivist philosophy of science has steered toward Charybdis in giving primacy to theory in conceptual understanding. Yet the implicit division of labor between conceptual development as theory-construction and its "external" empirical accountability blocks any passage between Scylla and Charybdis. Or so I shall argue.

EXPERIMENTATION AND THE DOUBLE MEDIATION OF THEORETICAL UNDERSTANDING

Philosophical work on scientific theories now often emphasizes the need for models to articulate their content. Morgan and Morrison (1999) influentially describe models as partially autonomous *mediators* between theories and the world. Theories do not confront the world directly but instead apply to models as relatively independent, abstract representations. Discussions of models as mediators have nevertheless attended more to relations between theories and models than to those between models and the world. In seeking to understand how science allows aspects of the world to show up within the space of reasons, I cannot settle for this starting point. Yet I must also avoid resorting to the Myth of the Given. Neither data nor other observable intermediaries are "Given" manifestations of the world.

My proposed path between Scylla and Charybis begins with Hacking's (1983) conception of "phenomena," which is quite different from Bogen and Woodward's use of the term for events-under-a-description. Phenomena in Hacking's sense are publicly accessible events in the world rather than linguistic or perceptual representations. He also more subtly shifts emphasis from what is observed or recognized to what is salient and noteworthy. Phenomena *show* something important about the world, rather than our merely *finding* something there. Hacking's term also clearly has a normative significance that cannot refer to something merely "given." Most events in nature or laboratories are not phenomena, for such events show little or nothing. Creating a phenomenon is an achievement, whose focus is the salience and clarity of a pattern against a background. Hacking suggested that

> old science on every continent [began] with the stars, because only the skies afford some phenomena on display, with many more that can be obtained by careful observation and collation. Only the planets, and more distant bodies,

have the right combination of complex regularity against a background of chaos. (1983, 227)[2]

Some natural events have the requisite salience and clarity, but most phenomena must be created.

I use Hacking's concept to respond to Morgan and Morrison that theoretical understanding is *doubly* mediated. "Phenomena" mediate in turn between models and the world to enable conceptual understanding. Hacking himself may share this thought. He concluded,

> In nature there is just complexity, which we are remarkably able to analyze. We do so by distinguishing, in the mind, numerous different laws. We also do so, by presenting, in the laboratory, pure, isolated phenomena. (1983, 226)

I think the "analysis" he had in mind was to make the world's complexity intelligible by articulating it conceptually. To take Hacking's suggestion seriously, however, we must understand how recognition or creation of phenomena could be a scientific "analysis," complementary to nomological representation. Elgin (1991) makes an instructive distinction between the properties an event merely instantiates and those it exemplifies. Turning a flashlight instantiates the constant velocity of light in different inertial reference frames, but the Michelson/Morley experiment exemplifies it. Similarly, homeotic mutants exemplify a modularity of development that normal limb or eye development merely instantiates.

Consider Elgin's example of the Michelson/Morley experiment. The interferometer apparatus lets a light beam tangential to the earth's motion *show* any difference between its velocity and that of a perpendicular beam. No such difference is manifest, but this display of the constant velocity of light is contextual in a way belying the comparative abstraction of the description. We can talk about the constant velocity of coincident light beams traveling in different directions relative to the earth's motion without mentioning the beam-splitters, mirrors, compensation plates, or detectors that enable the Michelson/Morley phenomenon. We often represent phenomena in this way, abstracting from the requisite apparatus, shielding, and other surrounding circumstances. Such decontextualized events are precisely what Bogen and Woodward or Friedman meant by "phenomena." Yet Hacking argued that such decontextualizing talk is importantly misleading. Using a different example, he claimed,

> The Hall effect does not exist outside of certain kinds of apparatus . . . That sounds paradoxical. Does not a current passing through a conductor, at right angles to a magnetic field, produce a potential, anywhere in nature? Yes and no. If anywhere in nature there is such an arrangement, with no intervening causes, then the Hall effect occurs. But nowhere outside the laboratory is there such a pure arrangement. (1983, 226)

The apparatus that produces and sustains such events in isolation is integral to the phenomenon, as much a *part* of the Hall effect as the conductor, the current, and the magnetic field. Yet the consequent material contextuality of phenomena might give us pause. Surely conceptual understanding must transcend such particularity to capture the generality of concepts. Perhaps we must *say* what a phenomenon shows and not merely show it to articulate the world conceptually. I shall argue, however, that the phenomena themselves, and not merely their verbal characterization, have conceptual significance pointing beyond themselves.

WHY NOT DETERMINE CONCEPTS BY THEIR EMPIRICAL SUCCESSES?

I begin with sustained critical attention to two important but flawed attempts to ascribe conceptual significance to the phenomena themselves rather than under a description. Both fail to pass between Scylla and Charybdis. Consider first Hacking's own account of relations between models and phenomena as "self-vindication," which emphasizes their stable co-evolution. "Self-vindication" occurs in laboratory sciences because

> the . . . systematic and topical theories that we retain, . . . are true to different phenomena and different data domains. Theories are not checked by comparison with a passive world . . . We [instead] invent devices that produce data and isolate or create phenomena, and a network of different levels of theory is true to these phenomena . . . Thus there evolves a curious tailor-made fit between our ideas, our apparatus, and our observations. A coherence theory of truth? No, a coherence theory of thought, action, materials, and marks. (Hacking 1992, 57–58)

Hacking rightly emphasizes mutual coadaptation of models and phenomena. We nevertheless cannot understand the empirically grounded conceptual content of models as a self-vindicating coadaptation with their data domain. Hacking's proposal steers directly into McDowell's Charybdis, rendering conceptual understanding empty in splendidly coherent isolation. Hacking envisages stable, coherent domains, as scientific claims gradually become almost irrefutable by limiting their application to well-defined phenomena they already fit. That exemplar of conceptual stability, geometrical optics, tellingly illustrates his claim:

> Geometrical optics takes no cognizance of the fact that all shadows have blurred edges. The fine structure of shadows requires an instrumentarium quite different from that of lenses and mirrors, together with a new systematic theory and topical hypotheses. Geometrical optics is true only to the phenomena of rectilinear propagation of light. Better: it is true of certain models of rectilinear propagation. (Hacking 1992, 55)

In supposedly securing the correctness of such theories within their domains, Hacking's proposal renders them empty. He helps himself to a presumption his own account undermines, namely that geometrical models of rectilinearity amount to a *theory* about *optics*. He indicates the difficulty in his concluding sentence: if the domain to be accounted for cannot be identified independent of the theory accounting for it, the theory is not *about* anything other than itself. It is one thing to say geometrical optics has limited effective range or only approximate accuracy. It is another thing to confine its domain to those phenomena that it accommodates. The fine structure of shadows *is* directly relevant to geometrical optics, and thereby displays the theory's empirical limitations. Only through such openness to empirical challenge does the theory purport to be *about* the propagation of light. McDowell's criticism of Davidson thus also applies to Hacking's view: he "manages to be comfortable with his coherentism, which dispenses with rational constraint upon [conceptual thought] from outside it, because he does not see that emptiness is the threat" (McDowell 1994, 68).

My second case is Nancy Cartwright's proposal that concepts like force in mechanics have limited scope. Force, she argues, is an abstract concept needing more concrete "fitting out." Just as I am not working unless I also do something more concrete like writing a paper, teaching a class, or thinking about the curriculum, so there is no force among the causes of a motion unless there is an approximately accurate force function such as $F = -kx$ or $F = mg$ (Cartwright 1999, 24–28, 37–46). Experimentation plays a role here, she argues, because these functions only apply accurately to "nomological machines" rather than to messier events. Cartwright's proposal does better than Hacking's, in allowing limited open-endedness to conceptual domains. The concept of force extends beyond the models for $F = ma$ actually in hand to apply wherever reasonably accurate models *could* be successfully developed.

This extension is still not sufficient. First, the domain of mechanics is then gerrymandered. Apparently similar situations, such as objects in free fall in the Earth's atmosphere, fall on different sides of its borders.[3] Second, this gerrymandered domain empties the concept of force of conceptual significance, and hence of content, because Cartwright conflates two dimensions of conceptual normativity.[4] A concept expresses a norm of classification with respect to which we may then succeed or fail to show how various circumstances accord with the norm.[5] Typically, we understand how and why it matters to apply *this* concept and group together these instances instead of or in addition to others. The difference it makes shows what is at stake in succeeding or failing to grasp things in accord with that concept (e.g., by finding an appropriate force function). Both dimensions of conceptual normativity are required: we need to specify the concept's domain (what it is a concept *of*) and to understand the difference between correct and incorrect application within that domain. By defining what is at stake in applying a concept *in terms of* criteria for its successful empirical application, she removes any meaningful stakes in that success. The concept then just *is* the classificatory grouping.[6] In removing a concept's accountability to an independently specifiable domain, Hacking and Cartwright thereby undermine both dimensions of conceptual normativity, because, as Wittgenstein (1953) famously noted, where one cannot talk about error one also cannot talk about correctness.

SALIENT PATTERNS AND CONCEPTUAL NORMATIVITY

Think again about phenomena in Hacking's sense. Their crucial feature is the manifestation of a meaningful pattern standing out against a background. This "standing out" need not be perceptual, of course. Some astronomical phenomena are visible to anyone who looks, but most require more effort. Experimental phenomena require actually arranging things to manifest a significant pattern, even if that pattern is subtle, elusive, or complex. As Karen Barad commented about a prominent recent case,

It is not trivial to detect the extant quantum behavior in quantum eraser experiments . . . In the quantum eraser experiment the interference pattern
was not evident if one only tracked the single detector [that was originally
sufficient to manifest a superposition in a two-slit apparatus] . . . What was
required to make the interference pattern evident upon the erasure of which-
path information was the tracking of two detectors simultaneously. (2007,
348–49)

That there is a pattern that stands out in an experimental phenomenon is
thus crucially linked to scientific capacities and skills for pattern recognition.
As Daniel Dennett once noted,

the self-contradictory air of "indiscernible pattern" should be taken
seriously . . . In the root case, a pattern is "by definition" a candidate for
pattern recognition. (1991, 32)

This link between "real patterns" and their recognition does not confer any
special privilege upon our capacities for discernment. Perhaps the pattern
shows up with complex instruments whose patterned output is discernible
only through sophisticated computer analysis of data. What *is* critical is the
normativity of recognition, to allow for the possibility of error. The patterns
that show up in phenomena must not merely indicate a psychological or
cultural propensity for responsiveness to them. Our responsiveness, *taking*
them as significant, must be open to assessment. What were once taken as
revealing patterns in the world have often been later rejected as misleading,
artifactual, or coincidental. The challenge is to understand how and why
those initially salient patterns lost their apparent significance and especially
why that loss corrects an earlier error rather than merely changing our de
facto responses.

Experimental phenomena are conceptually significant because the pattern they embody informatively refers beyond itself. To this extent, the salience of natural or experimental phenomena is broadly inductive.[7] Consider
the Morgan group's work initiating classical genetics. Their experiments
correlated differences in crossover frequencies of mutant traits with visible
differences in chromosomal cytology. If these correlations were peculiar to
Drosophila melanogaster, or worse, to these particular flies, they would have
had no scientific significance. Their salience instead indicated a more general

pattern in the cross-generational transmission of traits and the chromosomal location of "genes" as discrete causal factors.

Yet the philosophical issue here is not how to reason inductively from a telling instance of a concept to its wider applicability. We need to think about reflective judgment in the Kantian sense instead of the inductive-inferential acceptance of determinate judgments. The question concerns how to articulate and understand relevant conceptual content rather than how to justify judgments employing those concepts. The issue is a normative concern for how to articulate the phenomena understandingly, rather than a merely psychological consideration of how we arrive at one concept rather than another. In this respect, the issue descends from the "grue" problem of Goodman (1954). Goodman's concern was not why we actually project "green" rather than "grue," for which evolutionary and other considerations provide straightforward answers. His concern was why it is appropriate to project green, as opposed to why we (should) accept this or that judgment in either term.

Marc Lange's (2000) revisionist conception of natural laws helps here. For Lange, a law expresses how unexamined cases would behave in the same way as cases already considered. In taking a hypothesis to be a law, we commit ourselves to inductive strategies, and thus to the inductive projectibility of concepts employed in the law. Because many inference rules are consistent with any given body of data, Lange asks which possible inference rule is *salient*. The salient inference rule would impose neither artificial limitations upon its scope nor unmotivated bends in its further extension. Salience of an inference rule, Lange argues, is not

> something psychological, concerning the way our minds work . . . [Rather] it possesses a certain kind of justificatory status: in the manner characteristic of observation reports, this status [determines] . . . what would count as an unexamined [case] being relevantly the same as the [cases] already examined. (2000, 194)[8]

Whereas Lange compares salient rules to observation reports, however, I compare them to the salient pattern of a phenomenon. Its normative status as a salient pattern meaningfully articulates the world, rendering intelligible those aspects of the world falling within its scope, albeit defeasibly.

This role for meaningful patterns *in the world* does not steer back onto the philosophical rocks of Scylla. The salient patterns in natural or experimental phenomena are nothing Given but instead indicate the defeasibility of the pattern itself, and its scope and significance. One of Lange's examples illustrates this point especially clearly. Consider the pattern of correlated measurements of the pressure and volume of gases at constant temperature. Absent other considerations, their linear inverse correlation yields Boyle's law. Yet couple this same phenomenon with a model—one that identifies volume with the free space between gas molecules rather than the container size, and understands pressure as reduced by intermolecular forces diminishing rapidly with distance—and the salient pattern extension instead becomes the van der Waals law. How this pattern would continue "in the same way" at other volumes and pressures has shifted, such that the straight-line extension of Boyle's law incorporates an "unmotivated bend." Moreover, modeled and measured differently, all general patterns dissipate in favor of ones specific to the chemistry of each gas.

Recognizing the inherent normativity of pattern recognition in experimental practice recovers its requisite two dimensions. I criticized Hacking and Cartwright for defining the scope and content of scientific concepts by their successful applications. Yet they rightly looked to the back-and-forth between experimental phenomena and theoretical models for the articulation of conceptual content. Haugeland (1998) points us in the right direction by distinguishing

> two fundamentally different sorts of pattern recognition. On the one hand, there is recognizing an integral, present pattern from the outside—*outer recognition* . . . On the other hand, there is recognizing a global pattern from the inside, by recognizing whether what is present, the current element, fits the pattern— . . . *inner recognition*. The first is telling whether something (a pattern) is *there*; the second is telling whether what's there *belongs* (to a pattern). (1998, 285, italics in original)

A pattern is a candidate for outer recognition if what is salient in context points beyond itself in an informative way. The apparent pattern is not just an isolated curiosity or spurious association. Consequently, there is something genuinely at stake in how we extend this pattern, such that it can be done correctly or incorrectly. Only if it *matters* to distinguish those motions caused by forces from those that would not be so caused does classical mechanics have anything to be right or wrong about.[9] Whether a pattern indicates anything beyond its own occurrence is defeasible, in which case it shows itself to be a coincidental, merely apparent pattern.

Inner recognition identifies an element in or continuation of a larger pattern. Inner recognition is only at issue if some larger pattern is there, with something at stake in getting it right. Inner recognition grasps how to go on rightly, consonant with what is thereby at stake. For classical mechanics, inner recognition involves identifying forces and calculating their contributions to an outcome. But the existence of a pattern depends upon the possibility of recognizing how it applies. Haugeland thus concludes, rightly,

> What is crucial for [conceptual understanding] is that the two recognitive skills be distinct [even though mutually constitutive]. In particular, skillful practitioners must be able to find them in conflict—that is, simultaneously to outer-recognize some phenomenon as present (actual) and inner-recognize it as not allowed (impossible). (1998, 286)[10]

Both dimensions of conceptual normativity are needed to sustain the claim that the pattern apparently displayed in a phenomenon enhances the world's intelligibility. Something must be genuinely at stake in recognizing that pattern, and any issues that arise in tracking that pattern must be resolvable without betraying what was at stake.

MODELS AND CONCEPTUAL ARTICULATION

A two-dimensional account of the normativity of pattern recognition puts Cartwright's and Hacking's discussions of laboratory phenomena and theoretical modeling in a new light. Cartwright challenges this conception of scientific understanding in her work, from *How the Laws of Physics Lie* (1983) to *The Dappled World* (1999), by challenging the compatibility of inner and outer recognition in physics. In the first work, she argued that explanatory patterns expressed in fundamental laws and concepts are not candidates for inner recognition because most events in the world cannot be accurately treated in those terms without ad hoc phenomenological emendation and *ceteris paribus* hedging. In the latter work she argued that the alleged universality of the fundamental laws is illusory. The scope of their concepts is restricted to those situations, nomological machines, that actually generate more or less law-like behavior and the broader tendencies of their causal capacities. In the dappled world we live in, we need other, less precise concepts and laws. Scientific understanding is a patchwork rather than a conceptually unified field.

Cartwright draws upon two importantly connected features of scientific work. First, concepts that express the patterns projected inductively from revealing experimental or natural phenomena often outrun the relatively limited domains in which scientists understand their application in detail. Cartwright's examples typically involve mathematical theories where only a limited number of situations can be described and modeled accurately in terms of the theory, yet the point applies more generally. Classical genetics mapped phenotypic differences onto relative locations on chromosomes, but only a very few organisms were mapped sufficiently to allow genes correlated with traits to be localized in this way. Moreover, substantial practical barriers prevented establishing for most organisms the standardized breeding stocks and a wide enough range of recognized phenotypic mutations to allow for sufficiently dense and accurate mapping. Second, in "gaps" where one set of theoretical concepts cannot be applied in detail, other patterns often provide alternative ways of understanding and predicting behavior of interest. Cartwright cites the apparent overlap between classical mechanics and fluid mechanics (1999, chapter 1). The motion of a bank note in a swirling wind does not permit a well-defined force function for the causal effects of the wind, but the situation may well be more tractable in different terms. This issue has been widely discussed in one direction as reduction or supervenience relations between theoretical domains, but the relation goes in both directions: the supposedly supervening conceptual domain might be said to explicate the concepts or events that cannot be accurately modeled in terms of a "lower" or more basic level of analysis.[11]

Cartwright has identified important issues, but her response remains unsatisfactory. Her conclusion that "fundamental" physical concepts have limited scope depends upon a familiar but untenable account of grasping a concept and applying it. In this view, grasping a concept is (implicitly) grasping what it means in every possible, relevant situation. Here, Cartwright agrees with her "fundamentalist" opponents that $F = ma$, the quantum mechanical formalism, and other theoretical principles provide fully general schemata for applying their constituent concepts within their domains; she only disagrees about how far those domains extend. The fundamentalist takes their domain as unrestricted, with only epistemic limits on working out their application. Cartwright ascribes semantic and perhaps even metaphysical significance to those limitations, which she takes to display the inapplicability of those concepts.[12]

The alternative account by Wilson (2006) of empirical and mathematical concepts not only shows why this shared account of conceptual understanding is untenable, but also helps clarify how to acknowledge and respond to Cartwright's underlying concerns while also reconciling them with my concern for conceptual understanding. I want to understand the conceptual significance of experimental phenomena and their relation to practices of theoretical modeling without losing contact with what is at stake in applying scientific concepts. Cartwright thinks the dappled, patchwork character of the world is not amenable to smooth, systematic inclusion within the supposedly regimented universality of fundamental physical concepts. Wilson rejects the underlying "classical picture of concepts" and instead treats empirical concepts as organized in varying ways, such as loosely unified patchworks of facades bound together into atlases, or overlapping patchworks pulled in different directions by competing "directivities."[13] A fully general concept need not have any fully general way of applying it. As a telling example, he addresses

> the popular categorization of classical physics as *billiard ball mechanics*. In point of fact, it is quite unlikely that any treatment of the *fully generic* billiard ball collision can be found anywhere in the physical literature. Instead, one is usually provided with accounts that work approximately well in a limited range of cases, coupled with a footnote of the "for more details, see . . ." type . . . [These] specialist texts do not simply "add more details" to Newton, but commonly overturn the underpinnings of the older treatments altogether. (Wilson 2006, 180–81, italics in original)

The sequence of models treats billiard balls incompatibly first as point masses, then rigid bodies, almost-rigid bodies with corrections for energy loss, elastic solids distorting on impact, then with shock waves moving through the ball, generating explosive collisions at high velocities, and so on. Some models also break down the balls' impact into stages, each modeled differently, with gaps. Wilson concludes that "to the best I know, this lengthy chain of billiard ball declination never reaches bottom" (2006, 181).

Wilson provides extraordinarily rich case studies of disparate links among conceptual facades, patches, or platforms with accompanying "property dragging" that can shift how concepts apply in different settings. One distinction helps indicate the extent to which empirical concepts need not be smoothly regimented or fully determinately graspable. Think of sequences of billiard ball collision models as exemplifying an *intensifying* articulation of concepts with increasing precision and fine-grained detail. We then also need *extensive* articulation to adapt familiar concepts to unfamiliar circumstances. Wilson objects here to what he calls "tropospheric complacency":

> We readily fancy that we already "know what it is like" to be *red* or *solid* or *icy everywhere,* even in alien circumstances subject to violent gravitational tides or unimaginable temperatures, deep within the ground under extreme pressures, or at size scales much smaller or grander than our own, and so forth. (2006, 55, italics in original)

Thought experiments such as programming a machine to find rubies on Pluto (2006, 231–33) tellingly indicate the parochial character of confidence that we already know how to apply familiar concepts outside familiar settings (or even that the correct application is determinate).

Rejecting such complacency allows endorsement of Cartwright's denial that general law-schemata are sufficient to understand more complex or less accommodating settings, while rejecting limitations on the scope of their concepts. Concepts commit us to more than we know how to say or do. "Force" or "gene" should be understood as "dappled concepts" rather than as uniformly projectable concepts with limited scope in a dappled world. Brandom (1994, 583) suggests a telling analogy between conceptual understanding and grasping a stick. We may only firmly grasp a concept at one end of its domain, but we take hold of its entirety from that end. We are also accountable for unanticipated consequences of its use elsewhere. The same is true for pattern recognition in experimental work. These patterns can be inductively salient far beyond what we know how to say or act upon.

That open texture is why I discuss inner and outer recognition in terms of the issues and stakes in concept use. "Issues" and "stakes" are fundamentally anaphoric concepts. They allow reference to the scope and significance of a pattern, a concept, or a practice (what is at stake there), and what it would be for them to go on in the same way under other circumstances or more stringent demands (what is at issue), even though those issues and stakes might be contested or unknown. Recognizing the anaphoric character of conceptual normativity lets us see what is wrong with Lange's claim that inner recognition is shaped by disciplinary interests. He says,

A discipline's concerns affect what it takes for an inference rule to qualify as "reliable" there. They limit the error that can be tolerated in a certain prediction . . . as well as deem certain facts to be entirely outside the field's range of interests . . . With regard to a fact with which a discipline is *not* concerned, *any* inference rule is *trivially* accurate enough for that discipline's purposes. (2000, 228, italics in original)

Lange makes an important point here that is misleadingly expressed in terms of scientific disciplines and their concerns. First, what matters is not de facto interests of a discipline but what is at stake in its practices and achievements. Scientists can be wrong about what is at stake in their work, and those stakes can shift over time as the discipline develops. Second, the relevant locus of the stakes in empirical science is not disciplines as social institutions but domains of inquiry to which disciplines are accountable. The formation and maintenance of a scientific discipline is a commitment to the intelligibility and empirical accountability of a domain of inquiry with respect to its issues and stakes.

These considerations about conceptual normativity also refine Hacking's notion of phenomena as salient, informative patterns. The concepts developed to express what is inductively salient in a phenomenon are always open to further intensive and extensive articulation. The same is true of experimental phenomena. The implicit suggestion that phenomena are stable patterns of salience should give way to recognition of the interconnected dynamics of ongoing experimentation and model building.[14]

Thus far, I have discussed experimental phenomena as if experimenters merely established a significant pattern in the world, whose conceptual role needed further articulation by model building. That impression understates the conceptual significance of experimentation. Instead of distinct experimental phenomena, we should consider systematically interconnected experimental capacities. Salient patterns manifest in experimentation articulate whole domains of conceptual relationships rather than single concepts (Rouse 2008). Moreover, what matters is not a static experimental setting, but its ongoing differential reproduction, as new, potentially destabilizing elements are introduced into a relatively well-understood system. As Barad argues,

[Scientific] apparatuses are constituted through particular practices that are perpetually open to rearrangements, rearticulations, and other reworkings.

That is part of the creativity and difficulty of doing science: getting the instrumentation to work in a particular way for a particular purpose (which is always open to the possibility of being changed during the experiment as different insights are gained). (2007, 170)

The shifting dynamics of conceptual articulation in the differential reproduction of experimental systems suggests that all scientific concepts are dappled—that is, open to further intensive and extensive articulation in ways that might be only patchily linked together. That is not a deficiency. The supposed ideal of a completely articulated, accurate, and precise conceptual understanding is very far from ideal. Consider Lange's example of conceptual relations among pressure, temperature, and volume of gases. Neither Boyle's nor van der Waals's law yields a fully accurate, general characterization of these correlated macroproperties or the corresponding concepts. Yet each law brings a real pattern in the world to the fore, despite noise it cannot fully accommodate. These laws are not approximations to a more accurate but perhaps messy and complex relation among these macroproperties. Any treatment of pressure, temperature, and volume more precise than van der Waals's law requires attending to the chemical specificity of gases, and because gases can be mixed proportionally, the relevant variability has no obvious limit. Insisting upon more precise specification of these correlations abandons any generally applicable conceptual relationship among these properties, as these relationships *only* show up *ceteris paribus*.

Hacking was nevertheless right to recognize the stabilization of some scientific conceptual relationships, even if such stability is not self-vindicating. The patterns already disclosed and modeled in a scientific field are sometimes sufficiently articulated with respect to what is at stake in its inquiries. The situations where inner recognition of those conceptual patterns might falter if pushed far enough do not matter to scientific understanding, and those divergences are rightly set aside as noise. That is why Lange indexes natural laws to scientific disciplines, or better, to their domain-constitutive stakes. The scientific irrelevance of some gaps or breakdowns in theoretical understanding can hold even when more refined experimental systems or theoretical models are needed in engineering or other practical contexts.[15]

At other times, seemingly marginal phenomena, such as the fine-grained edges of shadows, the indistinguishable precipitation patterns of normal and cancerous cells in the ultracentrifuge, the discrete wavelengths of photoelectric emission, or subtle shifts in the kernel patterning of maize visible only to an extraordinarily skilled and prepared eye, matter in ways that conceptually reorganize a whole region of inquiry. That is why, contra Cartwright, the scope of scientific concepts extends further and deeper than their application can be accurately modeled, even when their current articulation seems sufficient to their scientific stakes.

A NEW SCIENTIFIC IMAGE?

I conclude with a provocative suggestion. Bas van Fraassen (1980) some years ago proposed a dramatic reconception of Sellars's account of the manifest and scientific images. Sellars (1963) argued that the explanatory power of the scientific image provided epistemic and metaphysical primacy over the manifest image of "humanity-in-the-world." Van Fraassen challenged Sellars's account via two central considerations. He first argued that explanatory power is only a pragmatic virtue of theories and not their fundamental accomplishment. Second, he proposed limiting belief in the scientific image to where it is rationally accountable to human observation, due to its privileged role in justifying *our* beliefs. His constructive empiricism thereby restored priority to the manifest image as source of epistemic norms.

My arguments suggest a different revision of the scientific image and its relation to our self-understanding. This revised image puts conceptual articulation at the center of the scientific enterprise rather than as a preliminary step toward justified beliefs or explanatory power. The sciences expand and reconfigure the space of reasons in both breadth and depth. At their best, they extend and clarify those aspects of the world that fall within the scope of what we can say, reason about, and act responsively and responsibly toward. The sciences do so in part by creating phenomena, extending them beyond the laboratory, and constructing and refining models that further articulate the world conceptually. That achievement often reconfigures the world itself to show itself differently and reorients ourselves to be responsive to new patterns that reconstitute what is at issue and at stake there. Conceptual understanding does require reasoned critical scrutiny of what we say and do, and holding performances and commitments accountable to evidence. Justification is more than just an optional virtue. Yet justification is always contextual, responsive to what is at issue and at stake in various circumstances. We therefore should not replace Sellars's emphasis upon explanation with van Fraassen's general conception of empirical adequacy as the telos of the scientific image. Empirical adequacy can be assessed at various levels of conceptual articulation, in response to different issues and concerns. Empirical adequacy is contextual and pragmatic, just as van Fraassen insisted about explanation.

The resulting reconception of the scientific image also revises again the relation between that image and our understanding of ourselves as persons accountable to norms. It yields a scientific conception of the world that would privilege neither a nature seemingly indifferent to normativity, nor a humanism that subordinates a scientific conception of the world to human capacities or interests. Karen Barad aptly expresses such a reconception of the Sellarsian clash of images in the title of her book: *Meeting the Universe Halfway* (2007). Working out how to meet the universe halfway—by grasping scientific and ethical understanding as part of scientifically articulated nature, and as responsive to issues and stakes that are not just up to us— requires a long story and another occasion.[16] Yet such a reconception of the scientific image is ultimately what is at stake in understanding the role of experimentation in conceptual understanding.

NOTES

1. McDowell (1984) actually invokes the figures of Scylla and Charybdis in a different context, namely what it is to follow a rule. There, Scylla is the notion that we can only follow rules by an explicit interpretation, and Charybdis is regularism, the notion that rule-following is just a blind, habitual regularity. I adapt the analogy to his later argument in McDowell (1994), with different parallels to Scylla and Charybdis, because the form of the argument is similar and the figures of Scylla and Charybdis are especially apt there. It is crucial not to confuse the two contexts, however; when McDowell (1984) talks about a "pattern," he means a pattern of behavior supposedly in accord with a rule, whereas when I talk about "patterns" later I mean the salient pattern of events in the world in a natural or experimental phenomenon.

2. "Once a genuine effect is achieved, that is enough. The [scientist] need not go on running the experiment again and again to lay bare a regularity before our eyes. A single case, if it is the right case, will do" (Cartwright 1989, 92). There are, admittedly, some phenomena for which "regularity" seems more appropriate, such as the recurrent patterns of the fixed stars or the robustness of normal morphological development. The phenomenon in

such cases is not the striking pattern of any of the constellations or the specific genetic, epigenetic, and morphological sequences through which tetrapod limbs develop; it is instead the robust regularity of their recurrence. In these cases, however, the regularity itself *is* the phenomenon rather than a repetition of it. Humeans presume that a conjunction of events must occur repeatedly to be intelligible to us. Hacking and Cartwright, by contrast, treat some regularities as themselves temporally extended single occurrences. Phenomena in this sense are indeed *repeatable* under the right circumstances. Their contribution to scientific understanding, however, does not depend upon their actual repetition.

3. Cartwright's account also requires further specification to understand which models count as successful extensions of the theory. Wilson (2006), for example, argues that many of the extensions of classical mechanics beyond its core applications to rigid bodies involve extensive "property-dragging," "representational lifts," and more or less ad hoc "physics avoidance." Cartwright might take many of these cases to exemplify "the claim that to get a good representative model whose targeted claims are true (or true enough) we very often have to produce models that are not models of the theory" (2008, 40).

4. My objection to Cartwright's view is subtly but importantly different from those offered by Kitcher (1999) and in a review by Winsberg et al. (2000). They each claim that her account of the scope of laws is vacuous because it allegedly reduces to "laws apply only where they do." Their objections turn upon a tacit commitment to a Humean conception of law in denying that she can specify the domain of mechanics without reference to Newton's laws. Because Cartwright allows for the intelligibility of singular causes, however, she can identify the domain of mechanics with those *causes* of motion that can be successfully modeled by differential equations for a force function. My objection below raises a different problem that arises even if one can identify causes of motion without reference to laws, and thus could specify (in terms of causes) where the domain of the laws of mechanics is supposed to reside in her account. My objection concerns how the concepts (e.g., force) applied within that domain acquire content; because the concepts are defined in terms of their success conditions, she has no resources for understanding what this success amounts to.

5. Failures to bring a concept to bear upon various circumstances within its domain have a potentially double-edged significance. Initially, if the concept is taken *prima facie* to have relevant applicability, then the fail-

ure to articulate *how* it applies in these circumstances marks a failure of understanding on the part of those who attempt the application. Sustained failure, or the reinforcement of that failure by inferences to it from other conceptual norms, may shift the significance of failure from a failure of understanding by concept-users to a failure of intelligibility on the side of the concept itself.

6. One can see the point in another way by recognizing that the scope-limited conception of "force" that Cartwright advances *would* have conceptual content if there were some further significant difference demarcated by the difference between those systems and those with a well-defined force function. Otherwise, the domain of "force" in her account would characterize something like "mathematically analyzable motions" in much the same way that Hacking reduces geometrical optics to models of rectilinearity (rather than of the rectilinear propagation of light).

7. The salience of a pattern encountered in a scientific phenomenon, then, should be sharply distinguished from the kind of formalism highlighted in Kant's (1987) account of judgments of the beautiful or the sublime (as opposed to the broader account of reflective judgment sketched in the First Introduction), or from any psychological account of how and why patterns attract our interest or appreciation in isolation.

8. I argue later, however, that the role Lange here assigns to de facto agreement among competent observers is not appropriate in light of his (and my) larger purposes.

9. In the case of classical mechanics, we conclude that there are *no* motions within its domain that are not caused by forces (although of course quantum mechanics does permit such displacements, such as in quantum tunneling). That inclusiveness does not trivialize the concept in the way that Cartwright's restriction of scope to its approximately accurate models does, precisely because of the defeasible coincidence between inner and outer recognition (as we will see later). There is (if classical mechanics is indeed a domain of genuine scientific understanding) a conceivable gap, what Haugeland (1998, chapter 13) calls an "excluded zone," between what we *could* recognize as a relevant cause of motion and what we *can* understand with the conceptual resources of classical mechanics. No actual situations belong within the excluded zone because such occurrences are impossible. Yet such impossibilities must be conceivable and even recognizable. Moreover, if such impossibilities occurred and could not be explained away or isolated as a relevant domain limitation (as we do with quantum discontinuities), then what seemed like salient patterns in the various phenomena

of classical mechanics would turn out to have been artifacts, curiosities, or other misunderstandings. On this conception, to say that a phenomenon belongs within the domain of the concept of force is not to say that we yet understand how to model it in those terms; it does make the concept ultimately accountable to that phenomenon and empirically limited in its domain to the extent that it cannot be applied to that phenomenon to the degree of accuracy called for in context.

10. Haugeland talks about what is crucial for "objectivity" rather than for conceptual understanding. Yet objectivity matters to Haugeland only as the standard for understanding. I have argued elsewhere that Haugeland's appeal to objectivity is misconstrued and that conceptual understanding should be accountable not to "objects" (even in Haugeland's quite general and formal sense) but to what is at issue and at stake in various practices and performances (Rouse 2002, chapters 7–9).

11. I use "explicate" here for any domain relations for which the possibility of reduction or supervenience might be raised, even where we might rightly conclude that the explicating domain does not supervene. Thus, mental concepts explicate the domain of organismal behavior for some organisms with sufficiently flexible responsive repertoires, even if mental concepts do not supervene upon physical states or non-mental biological functions.

12. Cartwright does not assign this role to de facto epistemic limitations that might merely reflect failures of imagination or effort. The concepts apply wherever more generally applicable models *could* be developed that enable the situations in question to be described with sufficient accuracy in their terms, without ad hoc emendation. As she once succinctly put the relevant criterion of generality, "It is no theory that needs a new Hamiltonian for each new physical circumstance" (Cartwright 1983, 139).

13. Wilson identifies the "classical picture of concepts" with three assumptions:

> (i) we can determinately compare different agents with respect to the degree to which they share "conceptual contents";
>
> (ii) that initially unclear "concepts" can be successively refined by "clear thinking" until their "contents" emerge as impeccably clear and well defined;
>
> (iii) that the truth-values of claims involving such clarified notions can be regarded as fixed irrespective of our limited abilities to check them (2006, 4–5).

He also identifies it differently as "classical gluing," whereby predicate and property are reliably attached directly or indirectly (e.g., indirectly via a theoretical web and its attendant "hazy holism").

14. Rouse (1996) argues for a shift of philosophical understanding from a static to a dynamic conception of epistemology. More recently (2009) I interpreted the account of conceptual normativity in Rouse (2002) as a nonequilibrium conceptual and epistemic dynamics. Brandom (2011) develops this analogy more extensively in interpreting of Wilson (2006) as offering accounts of the statics, kinematics, and dynamics of concepts.

15. Most of Wilson's examples are drawn from materials science, engineering, and applied physics with close attention to the behavior of actual materials. Brandom (2011) suggested that these domains exemplify the pressure put on concepts when routinely examined and developed by professionals through multiple iterations of a feedback cycle of extensions to new cases. He notes jurisprudence as a similar case, in which concepts such as contract or property with acceptable uses in political thought are under pressure from application in case law. These concepts are generally in good order for what is at stake in some contexts but are open to indefinitely extendable intensive and extensive articulation for other purposes.

16. I make a beginning toward working out such an understanding in Rouse (2002) and develop it further in Rouse (2015).

REFERENCES

Barad, Karen. 2007. *Meeting the Universe Halfway: Quantum Physics and the Entanglement of Matter and Meaning.* Durham, N.C.: Duke University Press.

Bogen, James, and James Woodward. 1988. "Saving the Phenomena." *Philosophical Review* 98: 303–52.

Brandom, Robert. 1994. *Making It Explicit: Reasoning, Representing and Discursive Commitment.* Cambridge, Mass.: Harvard University Press.

Brandom, Robert. 2011. "Platforms, Patchworks, and Parking Garages: Wilson's Account of Conceptual Fine-Structure in *Wandering Significance.*" *Philosophy and Phenomenological Research* 82: 183–201.

Cartwright, Nancy D. 1983. *How the Laws of Physics Lie.* Oxford: Oxford University Press.

Cartwright, Nancy D. 1989. *Nature's Capacities and Their Measurement.* Oxford: Oxford University Press.

Cartwright, Nancy D. 1999. *The Dappled World*. Cambridge: Cambridge University Press.

Cartwright, Nancy D. 2008. "Reply to Daniela Bailer-Jones." In *Nancy Cartwright's Philosophy of Science,* edited by S. Hartmann et al., 38–40. New York: Routledge.

Dennett, Daniel. 1991. "Real Patterns." *Journal of Philosophy* 89: 27–51.

Elgin, Catherine. 1991. "Understanding in Art and Science." In *Philosophy and the Arts,* Midwest Studies in Philosophy, vol. 16, edited by P. French, T. Uehling Jr. and H. Wettstein, 196–208. Notre Dame, Ind.: University of Notre Dame Press.

Friedman, Michael. 1974. "Explanation and Scientific Understanding." *Journal of Philosophy* 71: 5–19.

Friedman, Michael. 1999. *Reconsidering Logical Positivism.* Cambridge: Cambridge University Press.

Goodman, Nelson. 1954. *Fact, Fiction and Forecast.* Cambridge, Mass.: Harvard University Press.

Hacking, Ian. 1983. *Representing and Intervening.* Cambridge: Cambridge University Press.

Hacking, Ian. 1992. "The Self-Vindication of the Laboratory Sciences." In *Science as Practice and Culture*, edited by A. Pickering, 29–64. Chicago: University of Chicago Press.

Haugeland, John. 1998. *Having Thought.* Cambridge, Mass.: Harvard University Press.

Kant, Immanuel. 1987. *Critique of Judgment,* translated by W. Pluhar. Indianapolis: Hackett.

Kitcher, Philip. 1999. "Unification as a Regulative Ideal." *Perspectives on Science* 7: 337–48.

Lange, Marc. 2000. *Natural Laws in Scientific Practice.* Oxford: Oxford University Press.

McDowell, John. 1984. "Wittgenstein on Following a Rule." *Synthese* 58: 325–63.

McDowell, John. 1994. *Mind and World.* Cambridge, Mass.: Harvard University Press.

Morgan, Mary, and Margaret Morrison. 1999. *Models as Mediators.* Cambridge: Cambridge University Press.

Quine, Willard V. O. 1953. "Two Dogmas of Empiricism." In *From a Logical Point of View,* 20–46. Cambridge, Mass.: Harvard University Press.

Quine, Willard V. O. 1960. *Word and Object.* Cambridge, Mass.: MIT Press.

Richardson, Alan. 1998. *Carnap's Construction of the World.* Cambridge: Cambridge University Press.

Rouse, Joseph. 1996. *Engaging Science.* Ithaca, N.Y.: Cornell University Press.

Rouse, Joseph. 2002. *How Scientific Practices Matter.* Chicago: University of Chicago Press.

Rouse, Joseph. 2008. "Laboratory Fictions." In *Fictions in Science,* edited by M. Suarez, 37–55. New York: Routledge.

Rouse, Joseph. 2009. "Standpoint Theories Reconsidered." *Hypatia* 24: 200–9.

Rouse, Joseph T. 2015. *Articulating the World: Conceptual Understanding and the Scientific Image.* Chicago: University of Chicago Press.

Sellars, Wilfrid. 1963. "Philosophy and the Scientific Image of Man." In *Science, Perception and Reality,* 1–37. London: Routledge and Kegan Paul.

van Fraassen, Bas C. 1980. *The Scientific Image.* Oxford: Oxford University Press.

Wilson, Mark. 2006. *Wandering Significance.* Oxford: Oxford University Press.

Winsberg, Eric, Mathias Frisch, Karen Mcrikangas Darling, and Arthur Fine. 2000. "Review, Nancy Cartwright, *The Dappled World.*" *Journal of Philosophy* 97: 403–8.

Wittgenstein, Ludwig. 1953. *Philosophical Investigations.* London: Blackwell.

4

MODELING/EXPERIMENTATION
The Synthetic Strategy in the Study of Genetic Circuits

TARJA KNUUTTILA AND ANDREA LOETTGERS

In philosophical discussions, models have been located between theories and experiments, often as some sort of go-betweens facilitating the points of contact between the two. Although the relationship between models and theories may seem closer than the one between models and experimentation, there is a growing body of literature that focuses on the similarities (and differences) between modeling and experimentation. The central questions of this discussion have concerned the common characteristics shared by modeling and experimenting as well as the various ways in which the inferences licensed by them are justified (e.g., Morgan 2003; Morrison 2009; Parker 2009; Winsberg 2009). Even though at the level of scientific practice scientists usually have no problem of distinguishing models and simulations from experiments, it has turned out difficult to characterize philosophically their distinctive features. In what follows we will attempt to clarify these issues by taking a cue from Peschard (2012) and considering the relationship of modeling and experimentation through a case in which they are used "in a tandem configuration" as Peschard (2013) puts it. We will study a particular modeling practice in the field of synthetic biology, whereby scientists study genetic circuits through a combination of experiments, mathematical models and their simulations, and synthetic models.[1]

With synthetic models we refer, in this context, to engineered genetic circuits that are built from genetic material and implemented in a natural cell environment.[2] In the study of genetic circuits they mediate between mathematical modeling and experimentation. Thus, an inquiry into how and why researchers in the field of synthetic biology construct and use synthetic models seems illuminating as regards the distinct features of model-ing and experimentation.

Moreover, what makes the modeling practice of synthetic biology espe-cially interesting for studying modeling and experimentation is that there is often no division of labor in synthetic biology laboratories: the same scien-tists typically engage in mathematical and synthetic modeling, and often-times even in experimentation. Consequently, one might expect that there are good reasons why synthetic biologists proceed in such a combinational manner. We argue that the gap between mathematical modeling and experi-mentation gave rise to the construction of synthetic models that are built from genetic material using mathematical models as blueprints. These hy-brid entities share characteristics of both experiments and models, serving to make clearer the characteristic features of the two. Our empirical case is based, apart from the published journal articles in the field of synthetic bi-ology, on interviews conducted with some leading synthetic biologists and a laboratory study of research performed at the Elowitz Lab at the California Institute of Technology. Elowitz Lab is known for its cutting-edge study of genetic circuits.[3]

We will begin by reviewing the philosophical discussion of the relation-ship of modeling and experimentation (in the second section). After that we will consider the contributions of mathematical modeling, experimentation, and synthetic modeling to the study of gene regulation. We will focus, in particular, on two synthetic models and how their combination with math-ematical modeling gave researchers new insights that would have been unexpected on the basis of mathematical modeling alone (the third sec-tion). In the final section we discuss the characteristic features of modeling and experimentation, largely based on the previous discussion of the hybrid nature of synthetic models.

MODELING VERSUS EXPERIMENTATION

The experimental side of models as a distinct topic has largely emerged within the practice-oriented studies on modeling. Several philosophers have noted the experimental side of models, yet there has emerged no consensus on whether modeling and experimentation could or could not be clearly dis-tinguished from each other. While some philosophers think that there are crucial differences between modeling and experimentation in terms of their respective targets, epistemic results, or materiality, others have presented counterarguments that have largely run modeling/simulation and experimentation together. In the following we will review this discussion, raising especially those topics that are relevant for synthetic modeling.

The recent philosophical discussion has pointed out three ways in which modeling and simulation resemble each other. First of all, one can consider them as largely analogous operations aiming to *isolate* some core causal factors and their effects. This argument has been presented in the context of mathematical modeling. Second, in dealing with simulation, numerous philosophers and scientists alike have pointed out their experimental nature as kinds of "numerical experiments." The stress here is on *intervention* rather than on isolation. Also the fact that simulations are performed on a material device, the digital computer, has led some philosophers to liken them to experiments. A third motivation for claiming that the practices of modeling and experimentation are similar to each other invokes the fact that both simulationists and experimentalists produce data and are dealing with data analysis and error management (see Winsberg 2003, and Barberousse, Franceschelli, and Imbert 2009, for somewhat divergent views on this matter). We will not discuss this point further; rather, we will concentrate on isolation and intervention.

According to the argument from isolation, both in modeling and in experimentation one aims to *seal off* the influence of other causal factors in order to study how a causal factor operates on its own. In experimentation this sealing off happens through experimental controls, but modelers use various techniques such as abstraction, idealization, and omission as vehicles of isolation (see, e.g., Cartwright 1999; Mäki 2005). Consequently, a theoretical model can be considered as an outcome of the *method of isolation:* in modeling, a set of elements is theoretically removed from the influence of other elements through the use of a variety of unrealistic assumptions (Mäki 1992).

One central problem of the isolationist view is due to the fact that the idealizing and simplifying assumptions made in modeling are often driven by the requirements of tractability rather than those of isolation. The model assumptions do not merely neutralize the effect of the other causal factors but rather construct the modeled situation in such a way that it can be conveniently mathematically modeled (see, e.g., Cartwright 1999; Morrison 2008). This feature of mathematical models is further enhanced by their use of general, cross-disciplinary computational templates that are, in the modeling process, adjusted to fit the field of application (Humphreys 2004;

Knuuttila and Loettgers 2012, 2014, 2016a, 2016b). Such templates are often transferred from other disciplines, as in the case of the research on genetic circuits where many models, formal methods, and related concepts originate from physics and engineering (e.g., the concepts of oscillator, feedback mechanism, and noise, as we discuss later).

The second sense in which models and experiments may resemble each other is due to the fact that in both modeling and experimentation one seeks to *intervene on* a system in the light of the results of this intervention. Herein lies the idea that simulations are experiments performed on mathematical models. But the question is how deep this resemblance cuts. Two issues in particular have sparked discussion: the supposed target systems of simulations versus experiments, and the role of materiality they incorporate.

A common intuition seems to be that, whereas in experimentation one intervenes on the real target system of interest, in modeling one merely interacts with a model system (e.g., Gilbert and Troitzsch 1999; Barberousse, Franceschelli, and Imbert 2009). Yet a closer examination has assured several philosophers that these intuitions may be deceptive. Winsberg (2009) argues that both "experiments and simulations have objects on the one hand and targets on the other, and that, in each case, one has to argue that the object is suitable for studying the target" (579; see also Guala 2002). Thus, both experimentation and modeling/simulation seem to display features of surrogate reasoning (Swoyer 1991), which is visible, for instance, when one experiments with model organisms instead of the actual organisms of interest. Consequently, the relationship of a model or experiment to its respective target need not distinguish the two activities from each other. Peschard (2013) disagrees, however; she suggests that one should pay attention to the kinds of "target systems" that simulation and experimentation pick, respectively, *when they are used side by side* in scientific research. According to her, the target system of the simulation is the system represented by the manipulated model: "it is the system that simulation is designed to produce information about" (see also Giere 2009). In the case of experimentation, the target system is the experimental system, for example, the model organism. The experimental results concern the interventions on the model organism (e.g., rats), although the eventual *epistemic motivation* might be to gain information on the influence of a certain drug in humans.

Peschard's argument hints at the fact that at the level of scientific practice we often do not have any difficulties in distinguishing model systems from experimental systems, although borderline cases exist. Models and simulations are considered as kinds of representations even if they did not represent any real-world system but rather depicted a hypothetical or fictional system. The point is that being representations they typically are expressed in other media than from what their targets are made of, whereas experimental objects are supposed to share at least partly the same material makeup as the systems scientists are primarily interested in.

Indeed, the right kind of materiality has been claimed to be the distinguishing mark of experiments and even the reason for their epistemic superiority to simulations. The idea is that the relationship between a simulation and its target is nevertheless abstract, whereas the relationship between an experimental system and its target is grounded in the same material being supposedly governed by the same kinds of causes (Guala 2002; Morgan 2003). Consequently, while in simulation one experiments with a (formal) representation of the target system, in experimentation the experimental and target systems are made of the "same stuff." This difference also explains, according to Morgan and Guala, why experiments have more epistemic leverage than simulations. For example, anomalous experimental findings are more likely to incur change in our theoretical commitments than unexpected results from simulations (Morgan 2005).

Despite the intuitive appeal of the importance of the "same" materiality, it has been contested on different grounds. Morrison (2009) points out that even in the experimental contexts the causal connection with the physical systems of interest is often established via models (see, however, Giere 2009 for a counterargument). Consequently, materiality is not able to deliver an unequivocal epistemic standard that distinguishes simulation outputs from experimental results. Parker (2009) questions the alleged significance of the "same stuff." She interprets the "same stuff" to mean, for instance, the same fluid and points out that in traditional laboratory experiments on fluid phenomena many other things such as the depth of the fluid and the size, shape, roughness, and movement of any container holding it may matter. This leads her to suggest that it is the "relevant similarities" that matter for the justified inferences about the phenomena. Let us note that this renders, once again, modeling and experimentation close to each other.

In trying to get a firmer grip on the experimental nature of modeling and to which extent this perspective on modeling draws modeling and experimentation closer to each other, we will in the next sections take an excursion into the study of genetic circuits in the emerging field of synthetic biology. We will focus on the relationship of mathematical models, experiments on model organisms, and synthetic models in circadian clock research, which provides one of the most studied gene regulatory systems. All three activities are used in close combination within this field to study how genetic circuits regulate the rhythmic behavior of cells. This combinational approach has been portrayed by synthetic biologists as the "synthetic biology paradigm" (Sprinzak and Elowitz 2005). In the rest of this essay we discuss the epistemic rationale of this paradigm, which we call "synthetic strategy," and examine what it tells us about the characteristics of modeling and experimentation.

THE SYNTHETIC STRATEGY IN THE STUDY OF GENETIC CIRCUITS

Synthetic biology is a novel and highly interdisciplinary field located at the interface of engineering, physics, biology, chemistry, and mathematics. It developed along two main, though interdependent, paths: the engineering of novel biological components and systems, and the use of synthetic gene regulatory networks to gain understanding on the basic design principles underlying specific biological functions, such as the circadian clock. A particular modeling strategy, combinational modeling, is one of the distinctive features of the basic science approach to synthetic biology. It consists of a combinational use of model organisms, mathematical models, and synthetic models.

The basic idea of this combinational modeling strategy is shown in Figure 4.1, which is taken from a review article by Sprinzak and Elowitz (2005). The two authors call this approach "the synthetic biology paradigm." The diagram indicates two important differences between the *natural* and the *synthetic genetic circuits* (i.e., synthetic models):

1. The natural circuit exhibits a much higher degree of complexity than the synthetic circuit.
2. The synthetic circuit has been designed by using different genes and proteins than the natural circadian clock circuit.

Consequently, synthetic models have the advantage of being less complex than model organisms. On the other hand, in comparison with mathematical models they are of "the same materiality" as their biological counterparts, naturally evolved gene regulatory networks. That is, they consist of the same kind of interacting biochemical components and are embedded in

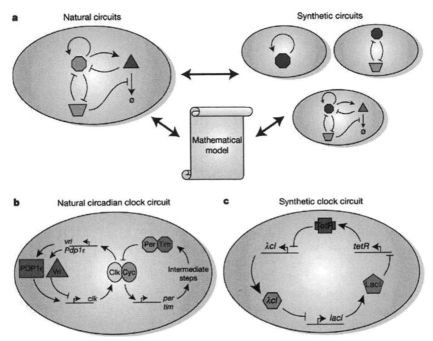

Figure 4.1. The "synthetic biology paradigm" according to Sprinzak and Elowitz (2005). The upper part of the diagram depicts the combinational modeling typical of synthetic strategy, and the lower part compares our present understanding of the circadian clock circuit of the fruit fly (*Drosophila melanogaster*) with a synthetic genetic circuit, the Repressilator.

a natural cell environment. This same materiality is crucial for the epistemic value of synthetic models: they are expected to operate in the same way as biological systems although the biochemical components are organized in a novel configuration.

In what follows, we argue that the combinational synthetic strategy arose in response to the constraints of mathematical modeling, on the one hand, and experimentation with model organisms on the other hand. There seemed to remain a gap between what could be studied mathematically and what scientists were able to establish experimentally. In particular, the experiments could not give any conclusive answer as to whether the hypothetical mechanisms that were suggested by mathematical models could be implemented as *molecular mechanisms capable of physically generating oscillatory phenomena* like the circadian rhythm. The problem was that in the experiments on model organisms such as *Drosophila* many genes and proteins underlying the circadian clock were found but the *dynamics of their interaction* regulating the day and night rhythm remained largely inaccessible. In consequence, the mechanism and dynamics suggested by mathematical models were probed by constructing synthetic models whose design was based on mathematical models and their simulations.

In the next sections, we will discuss in more detail the respective characteristics and specific constraints of mathematical modeling and the experimentation that gave rise to synthetic modeling. We start from mathematical modeling, which predated the experimental work in this area. Our discussion is based on one of the most basic models of gene regulation because it helps us to highlight some characteristic features of mathematical modeling that are less visible in subsequent models. Also, in pioneering works authors are usually more explicit about their modeling choices, and this certainly applies to Brian Goodwin's *Temporal Organization in Cells* (1963).

Mathematical Modeling

In *Temporal Organization in Cells* (1963) Brian Goodwin introduced one of the earliest and most influential mathematical models of the genetic circuit underlying the circadian rhythm. This book is an example of an attempt to apply concepts from engineering and physics to biology. Inspired by Jacob and Monod's (1961) operon model of gene regulation, Goodwin explored the mechanism underlying the temporal organization in biological systems, such as circadian rhythms, in terms of a negative feedback system. Another source of inspiration for him was the work of the physicist Edward Kerner (1957). Kerner had tried to formulate a statistical mechanics for the Lotka–Volterra model, which then prompted Goodwin to attempt to introduce a statistical mechanics for biological systems. These aims created the partly competing constraints of the design of Goodwin's model, both shaping its actual formulation and the way it was supposed to be understood. A third constraint was due to limitations of the mathematical tools for dealing with nonlinear dynamics. These three constraints are different in character. The first constraint was largely *conceptual,* but it had mathematical implications: the idea of a negative feedback mechanism that was borrowed from engineering provided the conceptual framework for Goodwin's work. It guided his conception of the possible mechanism and its mathematical form.

The second constraint, due primarily to the attempt to formulate a general theory according to the example provided by physics, was based on the assumption that biological systems should follow the basic laws of statistical mechanics. Goodwin described his approach in the following way: "The procedure of the present study is to discover conservation laws or invariants for a particular class of biochemical control systems, to construct a statistical mechanics for such a system, and to investigate the macroscopic behavior of the system in terms of variables of state analogous to those of physics: energy, temperature, entropy, free energy, etc." (1963, 7). This procedure provided a course of action but also imposed particular constraints such as the restriction to conservative systems, although biological systems are nonconservative as they exchange energy with the environment. We call this particular constraint, deriving from the goal of formulating a statistical mechanics for biological systems, a *fundamental* theory constraint. The third set of constraints was *mathematical* in character and related to the nature of the mathematical and computational methods used in the formulation of the model. Because of the nonlinearity introduced by the negative feedback loop, the tractability of the equations of the model posed a serious challenge.

The three types of constraints—conceptual, fundamental, and mathematical—characterize the theoretical toolbox available for mathematically modeling a specific system. But as a hammer reaches its limits when used as a screwdriver, the engineering concepts and especially the ideal of statistical mechanical explanation proved problematical when applied to biological systems. Indeed, in later systems and synthetic biology the aim for a statistical mechanics for biological systems was replaced by the search for possible *design principles* in biological systems. The concept of the negative feedback mechanism was preserved, functioning as a cornerstone for subsequent research, but there was still some uneasiness about it that eventually motivated the construction of synthetic models.

The toolbox-related constraints should be distinguished from the constraints more directly related to the biological systems to be modeled. These constraints are due to the enormously complex behavior of biological systems, which, as we will show, plays an important yet different role in mathematical modeling, experimentation, and synthetic modeling, respectively. Obviously the tool-related constraints are not static. Mathematical and computational methods develop, and theoretical concepts can change their meaning especially in interdisciplinary research contexts such as synthetic biology. Furthermore, the constraints are interdependent, and the task of a modeler consists in finding the right balance between them. We will exemplify these points by taking a closer look into the Goodwin model.

The basic structure of the network underlying the molecular mechanism of Goodwin's model of temporal organization is represented in Figure 4.2. The main structure of the model is rather simple, consisting of a negative feedback loop. It consists of a genetic locus L_i, synthesizing messenger RNA (mRNA) in quantities represented by the variable X_i. The mRNA leaves the nucleus and enters the ribosome, which reads out the information from the mRNA and synthesizes proteins in quantities denoted by Y_i. The proteins are connected to metabolic processes. At the cellular locus C the proteins influence a metabolic state, for example, by enzyme action, which results in the production of metabolic species in quantity M_i. A fraction of the metabolic species is traveling back to the genetic locus L_i where it represses the expression of the gene.

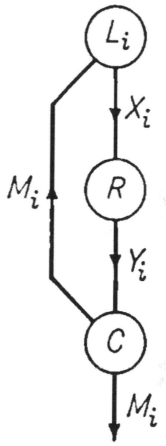

Figure 4.2. The circuit diagram underlying the Goodwin model (1963, 6).

This mechanism leads to oscillations in the protein level Y_i regulating temporal processes in the cell, such as the circadian rhythm. Goodwin described the mechanism by a set of coupled differential equations, which are nonlinear due to the feedback mechanism. The differential equations are of the following form:

$$\frac{dX_i}{dt} = a_i Y_i - b_i$$

$$\frac{dY_i}{dt} = c_i X_i - d_i$$

where $a_i Y_i$ describes the rate of mRNA controlled by the proteins synthetized and b_i its degradation. In the same way $c_i X_i$ describes the synthesis of the protein and d_i its degradation. The set of kinetic equations describes a deterministic dynamics.

In formulating his model, Goodwin had to make simplifying assumptions through which he attempted to deal with two important constraints: the complexity of the system consisting of a variety of biochemical components and processes, and the complexity due to the nonlinear dynamics of the assumed mechanism. First, he had to leave aside many known biochemical features of the circadian clock mechanism, and, second, he had to make assumptions that would allow him to simplify the mathematical model in such a way that he could use numerical methods to explore its complex dynamic behavior without having to solve the nonlinear coupled differential equations.

Goodwin was able to show by performing very basic computer simulations that the changes in the concentrations of mRNA and proteins form a closed trajectory. This means that the model system is able to perform regular oscillations, like those exhibited by circadian rhythms—but for the wrong reasons. Goodwin wrote, "The oscillations which have been demonstrated to occur in the dynamic system . . . persist only because of the absence of damping terms. This is characteristic of the behavior of conservative (integrable) systems, and it is associated with what has been called weak stability" (1963, 53). He went on to explain that a limit-cycle dynamics would have been the desirable dynamic behavior with respect to biological organisms. In this case, after small disturbances, the system moves back to its original trajectory—a characteristic of open (nonconservative) systems. But conceiving biological systems as open systems would have required the use of nonequilibrium statistical mechanics, which would have meant giving up some mathematical advantages of treating them as closed systems. Consequently, to simultaneously fulfill the aim of modeling a mechanism that produces oscillatory behavior *and* the aim to formulate a statistical mechanics for biological systems, Goodwin ended up presenting a model that he believed could only approximate the behavior he suspected to actually be taking place.

Even though Goodwin's model was not a complete success, it nevertheless provided a basic template for studying periodic biological phenomena at the molecular level. The subsequent modeling efforts were built on it, although in those endeavors the aim of formulating a statistical mechanics for biological systems was left behind. Instead, the notion of a feedback mechanism was made the centerpiece, the main constraints of the modeling approach now being the mathematical difficulties due to the nonlinear, coupled differential equations and the related problem of tractability. This meant that only some components and biochemical processes in natural systems could be taken into account. However, the subsequent models became more detailed as the first experiments exploring the molecular basis of the circadian rhythms became possible in the mid-1970s. These experiments, as we will show later, came with their very own constraints. Although modelers were able to partly bracket the complexity of biological systems by simply ignoring most of it, this was not possible in the experimental work.

Experimentation

During the 1960s the American biologist Colin Pittendrigh developed special methods and instruments that would allow him to study the circadian rhythms in rest and activity behavior and the eclosion of the fruit fly, *Drosophila melanogaster* (Pittendrigh 1954). *Drosophila* seemed to provide a good model organism for the study of the circadian rhythm. Take, for example, the eclosion. *Drosophila* goes through four different larva stages. At the end of the fourth larval stage the larva crawls into the soil and pupates. It takes the larva seven days to transform into a fly. When it emerges from the soil the outer layer of its body has to harden. At this stage the fly runs the risk of drying out. Because of this, its emergence from the soil should take place early in the morning when the air is still moist. Pittendrigh showed that the eclosion rhythm does not depend on any external cues such as light or temperature, and he reasoned that an endogenous mechanism—the circadian clock—is controlling the emergence of *Drosophila* from its pupa stage. Pittendrigh's studies provided the impetus for experiments in which scientists targeted the genes and mechanism underlying the circadian clock.

There were also other reasons for why *Drosophila* seemed to be the model organism of choice in isolating the genes contributing to the circadian clock. Before scientists started to use *Drosophila* in circadian clock research, it had already been extensively studied in genetics (see Kohler 1994). *Drosophila* breeds cheaply and reproduces quickly. Another important reason is its genetic tractability. The early genetic work suggested that the genome of the *Drosophila* contained only around 5,000 or 6,000 genes, no more than many bacteria. A further important incentive for choosing *Drosophila* is the size of its chromosomes. The salivary gland of the fruit fly contains giant chromosomes that can easily be studied under the microscope (see Morange 1998). Thus, when scientists started the investigation of the genetic makeup of the circadian clock, they had already at their disposal a lot of the experience and expertise from *Drosophila* genetics.

The first circadian clock gene of the *Drosophila* was found by Ronald Konopka and Seymour Benzer at the beginnings of the 1970s. They named the
gene *period* (*per*) because the mutants showed a drastic change in their normal 24-hour rhythm (Konopka and Benzer 1971). In their experiments Konopka and Benzer produced mutants by feeding chemicals to the flies. The
two scientists were especially interested in flies that showed changes in their
eclosion rhythm. The critical point of these experiments consisted of linking the deviation in the timing of the eclosion to the genetic makeup of the
fly. In this task Konopka and Benzer crossed male flies with a rhythm mutation with females carrying visible markers. The female and male off-
springs were mated again. The recombination of the genetic material in the
mothers led to various recombinants for the rhythm mutation. The visible
markers made it possible to locate genes linked to the circadian clock and to
identify the first circadian clock gene *period* (*per*).

The experimental research on circadian rhythms in molecular biology
and genetics progressed slowly after Konopka and Benzer published their
results. In the mid-1980s and early 1990s the situation started to change because of the further advances in molecular biology and genetics, such as
transgenic and gene knockout technologies. The transgenic methods make
use of recombinant DNA techniques to insert a foreign gene into the genome
of an organism, thus introducing new characters into it that were not present previously. Gene knockout technologies in turn inactivate specific genes
within an organism and determine the effect this has on the functioning of
the organism. As a result, more genes, proteins, and possible mechanisms
were discovered (see Bechtel 2011). However, despite the success of this experimental research program, the circadian clock research soon faced new
challenges as attention started to shift from isolating the clock genes and
their interrelations to the *dynamics* of these gene regulatory networks. To be
sure, the theoretical concepts already studied by the mathematical modeling approach, such as negative and positive feedback, had informed researchers in the piecemeal experimental work of identifying possible gene
regulatory mechanisms. For example, in mutation experiments different
possible mechanisms were tried out by varying the experimental parameters in systematic ways, and the results were related to such feedback mechanisms that could account for the observed behavior.

Although the experimental strategy in circadian clock research has thus
been very successful, it nevertheless has important constraints. First, there
are technical constraints related to the available methods from molecular

biology and genetics—such as creating the mutations and measuring the effects of the mutations. As already noted, the relaxation of some of these constraints led to the experimental approach that flourished from the mid-1980s onward. But even when it is possible to single out an explanatory mechanism experimentally, due to the complexity of biological systems there always remains the haunting question of whether all the components and their interactions have been found—whether the network is complete. This is a typical problem of the bottom-up experimental approach in molecular biology.

Another vexing question is whether the hypothetical mechanisms that researchers have proposed are the only ones that could explain the observations—which amounts to the traditional problem of underdetermination (which also besets the mathematical modeling approach to an even larger degree). But the most crucial constraint on the experimental approach—especially owing to the realization of the importance of the interaction between different genes and proteins—is the insufficient means of the experimental approach to study the dynamics. The experiments do not allow direct, but only indirect, observation of the network architecture and dynamics. As we will see, a new kind of modeling approach—synthetic modeling—was developed to *observe* the dynamics of genetic networks *within* the cell. This new modeling practice raised further important questions: how gene regulatory networks like the circadian clock interact with the rest of the cell, and how stochastic fluctuations in the number of proteins within the cell influence the behavior of the clock.

Synthetic Modeling

Although mathematical modeling (and simulation) and experimentation informed each other in circadian clock research—mathematical modeling suggesting possible mechanism designs, and experimentation in turn probing them and providing more biochemical detail—they still remained apart from each other. One important reason for this was that the modeling effort was based on rather schematic templates and related concepts, often originating from fields of inquiry other than biology. It was unclear whether biological organisms really functioned in the way the modelers suggested, even though the models were able to exhibit the kind of stable oscillations that were sought after.

The problem was twofold. First, the phenomena that the mathematical models were designed to account for could have been produced by other kinds of mechanisms. Second, and even more importantly, researchers were uncertain whether the general mechanisms suggested by mathematical models could be realized by biological organisms. The hybrid construction of synthetic models provides a way to deal with this problem. On the one hand, they are of the same materiality as natural gene regulatory networks because they are made of the same biological material, interacting genes and proteins. On the other hand, they differ from model organisms in that they are not the results of any evolutionary process, being instead designed on the basis of mathematical models. When engineering a synthetic model, a mathematical model is used as a kind of blueprint: it specifies the structure and dynamics giving rise to particular functions. Thus, the synthetic model has its origin in the mathematical model, but it is not bound by the same constraints. The model is constructed from the same material as the biological genetic networks, and it even works in the cell environment. Consequently, even if the synthetic model is not understood in all its details, it provides a simulation device of the same kind as its naturally evolved counterparts.

THE REPRESSILATOR

The Repressilator, an oscillatory genetic network, is one of the first and most famous synthetic models. It was introduced in 2000 by Michael Elowitz and Stanislas Leibler (2000). In an interview Michael Elowitz explained the basic motivation for constructing it:

> I was reading a lot of biology papers. And often what happens in those papers is that people do a set of experiments and they infer a set of regulatory interactions, and . . . at the end of the paper is a model, which tries to put together those regulatory interactions into a kind of cartoon with a bunch of arrows saying who is regulating who . . . And I think one of the frustrations . . . with those papers was [I was] always wondering whether indeed that particular configuration of arrows, that particular circuit diagram really was sufficient to produce the kinds of behaviors that were being studied. In other words, even if you infer that there's this arrow and the other arrow, *how do you know that that set of arrows is actually . . . working that way in the organism.* (Emphasis added)

Elowitz underlines in this quote the importance of studying the dynamics of network behavior, for which both mathematical and synthetic modeling provide apt tools. However, even though mathematical models can imitate the dynamics of biological phenomena, it is difficult to tell whether the kinds of networks that mathematical models depict really are implemented by nature. This is addressed by synthetic modeling.

From the perspective of the experimental features of modeling, it is important to notice how in the following statement Elowitz simultaneously portrays synthetic circuits as experiments and as a way of studying the general features of genetic circuits:

> It seemed like what we really wanted to do is to build these circuits to see what they really are doing inside the cell . . . That would be in a way the *best test of it* . . . of what kind of circuits were *sufficient for a particular function*. (Emphasis added).

The first step in constructing the Repressilator consisted in designing a mathematical model, which was used to explore the known basic biochemical parameters and their interactions. Next, having constructed a mathematical model of a gene regulatory network, Elowitz and Leibler performed computer simulations on the basis of the model. They showed that there were two possible types of solutions: "The system may converge toward a stable steady state, or the steady state may become unstable, leading to sustained limit-cycle oscillations" (Elowitz and Leibler 2000, 336). Furthermore, the numerical analysis of the model gave insights into the experimental parameters relevant for constructing the synthetic model and helped in choosing the three genes used in the design of the network.

The mathematical model functioned as a blueprint for engineering the biological system. The mathematical model is of the following form:

$$\frac{dm_i}{dt} = -m_i + \frac{\alpha}{(1+p_j^n)} + \alpha_0$$

$$\frac{dp_i}{dt} = -\beta(p_i - m_i)$$

with
$$\left(\begin{array}{l} i = lacI, tetR, cl \\ j = cl, lacI, tetR \end{array} \right)$$

In this set of equations, p_i is the concentration of the proteins suppressing the function of the neighbor genes, and m_i (where i is *lacI*, *tetR*, or *cl*) is the

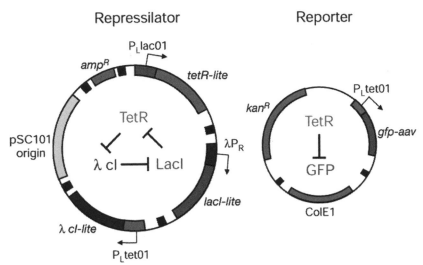

Figure 4.3. The main components of the Repressilator (left side) and the Reporter (right side). (Source: Elowitz and Leibler 2000, 336).

corresponding concentration of mRNA. All in all, there are six molecule species (three proteins functioning as repressors and three genes) all taking part in transcription, translation, and degradation reactions.

In Figure 4.3 the synthetic genetic regulatory network, the Repressilator, is shown on the left side and consists of two parts. The outer part is an illustration of the plasmid constructed by Elowitz and Leibler. The plasmid is an extrachromosomal DNA molecule integrating the three genes of the Repressilator. Plasmids occur naturally in bacteria. They can replicate independently, and in nature they carry genes that may enable the organism to survive, such as genes for antibiotic resistance. In the state of competence, bacteria are able to take up extrachromosomal DNA from the environment. In the case of the Repressilator, this property allowed the integration of the specific designed plasmid into *Escherichia coli* bacteria. The inner part of the illustration represents the dynamics between the three genes, *TetR*, *LacI*, and *λcI*. The three genes are connected by a negative feedback loop. This network design is directly taken from engineering, although the construction of it from molecular parts is a remarkable (bio)technological feat. Within electrical engineering this network design is known as the ring oscillator.

The left side of the diagram shows the *Reporter* consisting of a gene expressing green fluorescent protein (GFP), which is fused to one of the three genes of the Repressilator. The GFP oscillations in the protein level make visible the behavior of transformed cells, allowing researchers to study them over time by using fluorescence microscopy. This was quite epoch making as it enabled researchers to study the behavior of individual *living* cells. Green fluorescent proteins have been known since the 1960s but their utility as tools for molecular biology was recognized first in the 1990s, and several mutants of GFP were developed.[4] Apart from GFPs and plasmids, there were other important methods and technologies such as polymerase chain reaction (PCR) that enabled the construction of synthetic genetic circuits (Knuuttila and Loettgers 2013b).[5]

The Repressilator had only limited success. It was able to produce oscillations at the protein level, but these showed irregularities. From the perspective of the interplay of mathematical and synthetic modeling, the next move made by Elowitz and Leibler seems crucial: to find out what was causing such noisy behavior they reverted back to mathematical modeling. In designing the Repressilator, Elowitz and Leibler had used a deterministic model. A deterministic model does not take into account stochastic effects such as stochastic fluctuations in gene expression. Performing computer simulations on a stochastic version of the original mathematical model, Elowitz and Leibler were able to reproduce similar variations in the oscillations as observed in the synthetic model. This led researchers to conclude that stochastic effects may play a role in gene regulation—which gave rise to a new research program that targeted the role of noise in biological organization. This new research program revived the older discussions on nongenetic variation (e.g., Spudich and Koshland 1976), but this time the new tools such as GFP and synthetic modeling enabled researchers to observe nongenetic variation at an intracellular level. The researchers of the Elowitz Lab were especially interested in studying the various sources of noise in individual cells and the question of whether noise could be functional for living cells (e.g. Elowitz et al. 2002; Swain, Elowitz, and Siggia 2002).

The Dual-Feedback Synthetic Oscillator

The work of Jeff Hasty and his coworkers at University of California–San Diego provides another good example of how the combination of mathematical and synthetic modeling can lead to new insights that mere mathematical modeling (coupled with experimentation with model organisms) cannot provide. They constructed a dual-feedback synthetic oscillator exhibiting coupled positive and negative feedback, in which a promoter drives the production of both its own activator and repressor (Stricker et al. 2008; Cookson,

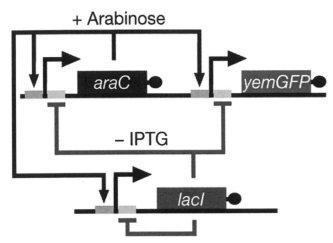

Figure 4.4. A diagrammatic representation of the dual-feedback oscillator (Source: Cookson, Tsimring, and Hasty 2009, 3934).

Tsimring, and Hasty 2009). The group introduced the synthetic oscillator by explaining how it was based on both experimental work on the *Drosophila melanogaster* clock and earlier theoretical work. The mathematical model providing the blueprint for the synthetic model was set forth by Hasty et al. (2002). This model, as Cookson et al. put it, was "theoretically capable of exhibiting periodic behavior" (Cookson, Tsimring, and Hasty 2009, 3932). Figure 4.4 shows the network diagram of the dual-feedback oscillator.

The synthetic network consists of two genes: *araC* and *lacI*. The hybrid promoter $P_{lac/ara-1}$ (the two small adjacent boxes) drives transcription of *araC* and *lacI*, forming positive and negative feedback loops. It is activated by the AraC protein in the presence of arabinose and repressed by the LacI protein in the absence of isopropyl β-D-1-thiogalactopyranoside (IPTG). The oscillations of the synthetic system are made visible by a GFP. Like the Repressilator, this system provides a material system allowing for the study of the oscillatory dynamic of the dual-feedback system; the basic properties and conditions of the oscillatory dynamic can be estimated on the basis of the components of the model and their interactions. When arabinose and IPTG are added to the system, the promoter becomes activated, and the system's two genes, *araC* and *lacI*, become transcribed. Increased production of the protein AraC in the presence of arabinose results in a positive feedback loop that increases the promoter activity. On the other hand, the increase in LacI production results in a linked negative feedback loop that decreases promoter activity. The difference in the promoter activities of the two feedback loops leads to the oscillatory behavior (Knuuttila & Loettgers 2013a, 2013b).

On the basis of the analysis of the dual-feedback oscillator, the researchers made several observations that were difficult to reconcile with the original mathematical model that had informed the construction of the synthetic model. The most surprising finding concerned the robustness of the oscillations. Whereas the mathematical model predicted robust oscillations only for a restricted set of parameters, the synthetic model instead showed robust oscillations for a much larger set of parameter values! The researchers were perplexed by the fact that "it was difficult to find inducer levels at which the system *did not* oscillate" (Cookson, Tsimring, and Hasty 2009, 3934, italics in original). They had anticipated that they would need to tune the synthetic model very carefully for it to yield oscillations, as the analysis of the mathematical model had shown that there was only a very small region in the parameter space where it would oscillate.

These contradictory results led the researchers in question to reconsider the original model. Cookson, Tsimring, and Hasty wrote about this as follows: "In other words, it became increasingly clear that the observed oscillations did not necessarily validate the model, even though the model predicted oscillations. We were able to resolve the discrepancy between the model and theory by reevaluating the assumptions that led to the derivation of the model equations" (2009, 3934). This statement does not reveal how laborious the actual modeling process was—it took a whole year. In the beginning the researchers had no clue how to account for the too-good performance of the model, or as the leader of the laboratory, Jeff Hasty, put it in an interview: "We did not understand why it was oscillating even though we designed it to oscillate." So he suggested to one of the postdoctoral researchers in the laboratory ("a modeling guy") an unusual way to proceed:

> We have sort of physicists' mentality: we like simple models. But [I thought that] maybe in this case we should start with just some big model where we model more of the interactions, and not try to simplify it into something eloquent . . . And, my postdoc was a bit reluctant because we did not know . . . many of the parameters, in anything we were doing . . . So he said: "What's the point of that?" . . . But my reasoning was, well, it should be robustly oscillating, so it should not really matter what the parameters are. You should see oscillations all over the place.[6]

With the bigger model the researchers were indeed able to create very robust oscillations. But the new problem was that "the big model was not necessarily telling us more than the big experiment."

However, when the researchers started to look at the time series of the model they found out that a small time delay in the negative feedback loop caused by a cascade of cellular processes was crucial for producing robust oscillations. Another unexpected finding concerned the enzymatic decay process of the proteins that turned out to be crucial for the sustained oscillations. The limited availability of the degradation machinery provided by the cell improved the precision of the oscillator by causing an indirect post-translational coupling between the activator (AraC) and repressor (LacI) proteins.[7] As such, this was a consequence of an unintended interaction with the host cell that showed that interactions with the cell environment can be advantageous for the functioning of synthetic circuits (for more details, see Knuuttila and Loettgers 2013a, 2013b). This is as remarkable an insight as the one concerning the functional role of noise, given that the original program of synthetic biology was based on the idea of constructing modular synthetic circuits from well-characterized components. Other researchers in the field of synthetic biology were quick to pick up the findings of Hasty and his coworkers, suggesting strategies to develop next-generation synthetic models that would integrate more closely with the endogenous cellular processes of the host organisms (Nandagopal and Elowitz 2011).

In conclusion, synthetic models such as the Repressilator and the dual-feedback oscillator have so far been limited to simple networks that are designed in view of what would be optimal for the behavior under study. This means that such networks need not be part of any naturally occurring system; they are chosen and tuned on the basis of the simulations of the underlying mathematical model and other background knowledge in such a way that the resulting mechanism would allow for (stable) oscillations. These technical constraints imply a constraint on what can be explored by such synthetic models: possible design principles in biological systems. Synthetic models are thus like mathematical models, but they take one step closer to empirical reality in showing that the theoretical mechanism in question is actually realizable in biological systems.

HOW TO UNDERSTAND THE EXPERIMENTAL
CHARACTER OF MODELING?

In the previous sections we studied the synthetic strategy in the study of genetic circuits, which is based on a close triangulation of experiments and mathematical and synthetic models. We have shown, in particular, how synthetic models have generated new theoretical insights as a result of their mismatch with the original mathematical models that provided the blueprints for their construction. As synthetic modeling involves a step toward materially embodied experimental systems, it seems instructive to study the philosophical discussion on modeling and experimentation through this case. In the second section we discussed two arguments for seeing models as kinds of experiments.

The argument from isolation draws an analogy to experimental practice in claiming that models are abstracted from real-world systems through suitable assumptions in order to study how a causal factor operates on its own. This does not suit the experimental practice of synthetic biology for reasons that seem to us applicable to many other modeling practices as well. First, instead of targeting the influence of isolated causal factors, the modelers in the field of systems and synthetic biology focus on the interaction of the (molecular) components of the assumed mechanism. What is at stake is whether the kinds of feedback loops that have extensively been studied in physics and engineering could generate the periodic/oscillatory behavior typical of biological systems. The issue is of great importance: in human engineered systems oscillations are predominantly a nuisance, but in biological systems they comprise one of the very ways through which they regulate themselves (Knuuttila and Loettgers 2013a, 2014). This was already pointed out by Brian Goodwin, who studied, in addition to circadian clocks, other biological feedback mechanisms such as metabolic cycles (Goodwin 1963).

William Bechtel and Adele Abrahamsen have argued, on the basis of a meticulous study of circadian clock research, that models provide dynamic mechanistic explanations. The idea is that while the experimental work typically aims to discover the basic components and operations of biological mechanisms (i.e., genes and associated proteins) the task of modeling is to study their organization and "dynamical orchestration" in time (e.g., Bechtel 2011; Bechtel and Abrahamsen 2011).[8] More generally, mathematical models in many sciences can be understood as systems of interdependencies, whose outcomes over time are studied by modeling. The reason mathematical modeling provides such an indispensable tool for science is that a human mind cannot track the interaction of several variables—and when it comes to nonlinear dynamics typical of feedback loops, computer simulations are needed even in the case of such simple systems as the Lotka–Volterra equations (Knuuttila and Loettgers 2012).

One conceivable way to rescue the idea of isolation would be to reformulate it in such a way that the focus is not on the behavior and outcomes of isolated causal factors but on isolating a mechanism instead. As synthetic biologists try to uncover what they call "the basic design principles" of biological organization, this argument may seem initially plausible. Yet from the perspective of the actual modeling heuristic of synthetic biology, it is questionable whether the idea of isolating a mechanism has any more than metaphorical force. As most of the biochemical details of the systems studied are unknown, the modeling process in this field cannot be understood as isolating a possible mechanism from empirically known components and their interactions.

As one of the pioneers of synthetic modeling described his situation at the end of 1990s, "I did not know what equations to write down. There was really not any literature that would guide me."[9] The modeling process that ensued from the surprisingly robust oscillations of the dual-feedback oscillator provides a nice example of this difficulty. Even though the researchers tried to add details to their original model, it did not recapitulate what they saw happening in their synthetic model. Only through using a larger, partly speculative model were the researchers able to arrive at some hypotheses guiding the consequent model building. Thus, the modeling process was more hypothetical in nature than it would have been had it primarily consisted of the abstracting of a mechanism from known empirical details—and then subsequently adding them back.

The other argument for treating models as experiments underlines the interventional dimension of modeling, which is especially clear in the case of computer simulations. In dealing with both computer simulations and experiments one usually intervenes in more complex and epistemically opaque systems than when one manipulates highly idealized, mathematically tractable models. Lenhard (2006) suggests that simulation methods produce "understanding by control," which is geared toward design rules and predictions. This provides an interesting analogy between simulations and experiments on model organisms: in both of them one attempts through control to overcome the complexity barrier. Although living organisms are radically more complex than simulation models, this feature of theirs can also be turned into an epistemic advantage: if one experiments with living organisms instead of *in silico* representations of them, the assumedly same material offers a causal handle on their behavior, as Guala and Morgan suggested.

Waters (2012) has lucidly spelled out the advantages of using material experimental models by pointing out that this strategy "avoids having to understand the details of the complexity, not by assuming that complexity is irrelevant but by incorporating the complexity in the models." We have seen how the epistemic rationale of synthetic modeling depends critically on this point. In constructing synthetic models from biological material and implementing them in the actual cell environment, researchers were able to discover new dimensions of the complex behavior of cells.

Yet synthetic models also go beyond the strategy of using model organisms in that they are more artificial, insofar as being novel, man-made biological entities and systems that are not the results of evolutionary processes. Aside from their use in developing useful applications, the artificial construction of synthetic models also can be utilized for theoretical purposes—and it makes them more akin to mathematical models than to experiments. Take the case of the Repressilator. It is less complex than any naturally occurring genetic circuit that produces circadian cyclic behavior. It is built of only three genes and a negative feedback loop between them. The three genes do not occur in such a constellation in any known genetic circuit, and even their order was carefully designed. Thus it was not supposed to imitate any natural circuit, but rather to identify the *sufficient components* and *interactions* of a mechanism able to produce a specific behavior, such as oscillations in protein levels. The components of the network were chosen to achieve the most optimal behavior, not to get as close as possible to any naturally evolved genetic network. This serves to show that what the researchers were after was to explore whether the various feedback systems studied in engineering and complex systems theory could create the kind of periodic self-regulatory behavior typical of biological organisms. Thus, synthetic modeling and mathematical modeling in biology are hypothesis-driven activities that seek to explore various theoretical possibilities.

It is important to note, however, that the model status of the Repressilator is not due to its being a model in the sense of being a representation of some real-world target system (cf. Weber 2012). It is difficult to see what it could be a model of, given its artificial constructed nature. It is rather a fiction—albeit a concrete one—made of biological material and implementing a theoretical mechanism. What makes it a model is its simplified, systemic, self-contained construction that is designed to explore certain pertinent theoretical questions. It is constructed in view of such questions and the outcomes or behavior that it is expected to exhibit. In this sense it is best conceived of as an intricate epistemic tool (for the notion of models as epistemic tools, see Knuuttila 2011).

In addition to the characteristics that synthetic circuits share with models, their specific epistemic value is due to the way these model-like features intertwine with their experimental nature. This is the reason why we consider the notion of an experimental model especially apt in the case of synthetic models. Several of the synthetic biologists we interviewed had a background in neurosciences, and they pointed out their dissatisfaction with the lack of empirical grounding in neuroscientific models. As discussed earlier, synthetic models are constructed to test whether the hypothetical mechanisms that were suggested by mathematical models could be implemented as molecular mechanisms capable of physically generating oscillatory molecular phenomena like the circadian rhythm. Thus, synthetic models derive their experimental empirical character from the fact that they are constructed from the same kinds of material components as their natural counterparts. They are best conceived of as engineered "natural" or "scientific kinds," giving more control to the researchers of the system under study than the domesticated "natural kinds" (such as model organisms).

Moreover, by being implemented in a natural cell environment synthetic models are exposed to some further constraints of naturally evolved biological systems. Although these constraints are in general not known in all their details, the fact that they are there is considered particularly useful in the contexts in which one has an imperfect understanding of the causal mechanisms at work. This explains how the researchers reacted to the unexpected results. In the case of experimentation, anomalous or unexpected results are commonly taken more seriously than in modeling—if the model does not produce what is expected of it, modelers usually try to devise a better model. This was how the researchers also proceeded in the cases of the Repressilator and the dual-feedback oscillator. They went back to reconfigure their original mathematical models and even constructed new ones in the light of the behavior of synthetic models.

WE HAVE ARGUED through the case of synthetic modeling that although modeling and experimentation come close to each other in several respects, they nevertheless have some quite distinct features and uses. This explains why they typically are, in cases where it is possible, triangulated in actual scientific practices. Occasionally, from the combined effort of modeling and experimentation new hybrids are born—such as synthetic genetic circuits— that may properly be called experimental models. A closer analysis of the hybrid nature of synthetic models serves to show what is characteristic about modeling and experimentation, respectively.

Synthetic models share with mathematical models their highly simpli-fied, self-contained construction, which makes them suitable for the explo-ration of various theoretical possibilities. Moreover, both mathematical models and synthetic models are valued for their systemic, dynamic fea-tures. On the other hand, synthetic models are like experiments in that they are expected to have the same material and causal makeup as the systems under study—which makes them more complex and opaque than mathemat-ical models, yielding surprising, often counterintuitive results. This feature that synthetic models share with experiments typically requires mathemati-cal modeling to find out what causes the behavior of synthetic circuits. The specific affordance of mathematical modeling in this task derives partly from its epistemic and quite literal "cheapness," as mathematical models are fairly easy to set up and reconfigure in contrast to experiments. The recent modeling practice of synthetic biology shows rather conclusively how the triangulation of mathematical and synthetic modeling has led to new in-sights that otherwise would have been difficult to generate by either math-ematical modeling or by way of experimentation on model organisms alone.

NOTES

1. Genetic circuits are also called gene regulatory networks.

2. In addition to genetic circuits, synthetic biology also studies meta-bolic and signaling networks.

3. The interview material is from an interview with the synthetic biologist Michael Elowitz that was conducted in April 2010 by Andrea Loettgers.

4. Martin Chalfie, Osamu Shimomura, and Roger Y. Tsien shared the 2008 Nobel Prize in Chemistry for their work on green fluorescent proteins.

5. Polymerase chain reaction is a molecular biology technique origi-nally introduced in 1983 that is now commonly used in medical and biologi-cal research laboratories to generate an exponentially increasing number of copies of a particular DNA sequence.

6. The interview material is from an interview with the synthetic biologist Jeff Hasty that was conducted in August 2013 by Tarja Knuuttila.

7. This limited availability of the degradation machinery led Hasty and his coworkers to study queuing theory in an attempt to mathematically model the enzymatic decay process in the cell (Mather et al. 2011).

8. For a partial critique, see Knuuttila and Loettgers 2013b. Bechtel and Abrahamsen assume in mechanist fashion that the experimentally decomposed components could be recomposed in modeling.

9. In such situations scientists often make use of model templates and concepts borrowed from other disciplines and subjects—in the case of synthetic biology, the modelers' toolbox includes engineering concepts such as feedback systems and modeling methods from nonlinear dynamics (see the discussion in the chapter).

REFERENCES

Barberousse, Anouk, Sara Franceschelli, and Cyrille Imbert. 2009. "Computer Simulations as Experiments." *Synthese* 169: 557–74.

Bechtel, William. 2011. "Mechanism and Biological Explanation." *Philosophy of Science* 78: 533–57.

Bechtel, William, and Adele Abrahamsen. 2011. "Complex Biological Mechanisms: Cyclic, Oscillatory, and Autonomous." In *Philosophy of Complex Systems,* Handbook of the Philosophy of Science, vol. 10, edited by C. A. Hooker, 257–85. Oxford: Elsevier.

Cartwright, Nancy. 1999. "The Vanity of Rigour in Economics: Theoretical Models and Galilean Experiments." Discussion Paper Series 43/99, Centre for Philosophy of Natural and Social Sciences, London School of Economics.

Cookson, Natalia A., Lev S. Tsimring, and Jeff Hasty. 2009. "The Pedestrian Watchmaker: Genetic Clocks from Engineered Oscillators." *FEBS Letters* 583: 3931–37.

Elowitz, Michael B., and Stanislas Leibler. 2000. "A Synthetic Oscillatory Network of Transcriptional Regulators." *Nature* 403: 335–38.

Elowitz, Michael B., Arnold J. Levine, Eric D. Siggia, and Peter S. Swain. 2002. "Stochastic Gene Expression in a Single Cell." *Science* 297: 1183–86.

Giere, Ronald N. 2009. "Is Computer Simulation Changing the Face of Experimentation?" *Philosophical Studies* 143: 59–62.

Gilbert, Nigel, and Klaus Troitzsch. 1999. *Simulation for the Social Scientists.* Philadelphia: Open University Press.

Goodwin, Brian C. 1963. *Temporal Organization in Cells: A Dynamic Theory of Cellular Control Processes*. New York: Academic Press.

Guala, Francesco. 2002. "Models, Simulations, and Experiments." In *Model-Based Reasoning: Science, Technology, Values*, edited by L. Magnani and N. Nersessian, 59–74. New York: Kluwer.

Hasty, Jeff, Milos Dolnik, Vivi Rottschäfer, and James J. Collins. 2002. "Synthetic Gene Network for Entraining and Amplifying Cellular Oscillations." *Physical Review Letters* 88: 148101.

Humphreys, Paul. 2004. *Extending Ourselves: Computational Science, Empiricism, and Scientific Method*. Oxford: Oxford University Press.

Jacob, François, and Jacques Monod. 1961. "Genetic Regulatory Mechanisms in the Synthesis of Proteins." *Journal of Molecular Biology* 3: 318–56.

Kerner, Edward H. 1957. "A Statistical Mechanics of Interacting Biological Species." *Bulletin of Mathematical Biophysics* 19: 121–46.

Knuuttila, Tarja. 2011. "Modeling and Representing: An Artefactual Approach." *Studies in History and Philosophy of Science* 42: 262–71.

Knuuttila, Tarja, and Andrea Loettgers. 2012. "The Productive Tension: Mechanisms vs. Templates in Modeling the Phenomena." In *Representations, Models, and Simulations*, edited by P. Humphreys and C. Imbert, 3–24. New York: Routledge.

Knuuttila, Tarja, and Andrea Loettgers. 2013a. "Basic Science through Engineering: Synthetic Modeling and the Idea of Biology-Inspired Engineering." *Studies in History and Philosophy of Science Part C: Studies in History and Philosophy of Biological and Biomedical Sciences* 44: 158–69.

Knuuttila, Tarja, and Andrea Loettgers. 2013b. "Synthetic Modeling and the Mechanistic Account: Material Recombination and Beyond." *Philosophy of Science* 80: 874–85.

Knuuttila, Tarja, and Andrea Loettgers. 2014. "Varieties of Noise: Analogical Reasoning in Synthetic Biology." *Studies in History and Philosophy of Science Part A* 48: 76–88.

Knuuttila, Tarja, and Andrea Loettgers. 2016a. "Modelling as Indirect Representation? The Lotka–Volterra Model Revisited." *British Journal for the Philosophy of Science* 68: 1007–36.

Knuuttila, Tarja, and Andrea Loettgers. 2016b. "Model Templates within and between Disciplines: From Magnets to Gases—and Socio-economic Systems." *European Journal for Philosophy of Science* 6: 377–400.

Kohler, E. Robert. 1994. *Lords of the Fly:* Drosophila *Genetics and the Experimental Life*. Chicago: University of Chicago Press.

Konopka, Ronald J., and Seymor Benzer. 1971. "Clock Mutants of *Drosophila melanogaster.*" *Proceedings of the National Academy of Sciences of the United States of America* 68: 2112–16.

Lenhard, Johannes. 2006. "Surprised by Nanowire: Simulation, Control, and Understanding." *Philosophy of Science* 73: 605–16.

Lenhard, Johannes. 2007. "Computer Simulation: The Cooperation between Experimenting and Modeling." *Philosophy of Science* 74: 176–94.

Mäki, Uskali. 1992. "On the Method of Isolation in Economics." *Poznań Studies in the Philosophy of Science and Humanities* 26: 316–51.

Mäki, Uskali. 2005. "Models Are Experiments, Experiments Are Models." *Journal of Economic Methodology* 12: 303–15.

Mather, W. H., J. Hasty, L. S. Tsimrig, and R. J. Williams. 2011. "Factorized Time-Dependent Distributions for Certain Multiclass Queueing Networks and an Application to Enzymatic Processing Networks." *Queuing Systems* 69: 313–28.

Morange, Michel. 1998. *A History of Molecular Biology.* Cambridge, Mass.: Harvard University Press.

Morgan, Mary S. 2003. "Experiments without Material Intervention: Model Experiments, Virtual Experiments and Virtually Experiments." In *The Philosophy of Scientific Experimentation,* edited by H. Radder, 216–35. Pittsburgh: University of Pittsburgh Press.

Morgan, Mary S. 2005. "Experiments versus Models: New Phenomena, Inference and Surprise." *Journal of Economic Methodology* 12: 317–29.

Morrison, Margaret. 2008. "Fictions, Representations, and Reality." In *Fictions in Science: Philosophical Essays on Modeling and Idealization,* edited by M. Suárez, 110–35. New York: Routledge.

Morrison, Margaret. 2009. "Models, Measurement and Computer Simulations: The Changing Face of Experimentation." *Philosophical Studies* 143: 33–57.

Nandagopal, Nagarajan, and Michael B. Elowitz. 2011. "Synthetic Biology: Integrated Gene Circuits." *Science* 333: 1244–48.

Parker, Wendy. 2009. "Does Matter Really Matter? Computer Simulations, Experiments and Materiality." *Synthese* 169: 483–96.

Peschard, Isabelle. 2012. "Forging Model/World Relations: Relevance and Reliability." *Philosophy of Science* 79: 749–60.

Peschard, Isabelle. 2013. "Les Simulations sont-elles de réels substituts de l'expérience?" In *Modéliser & simuler: Epistémologies et pratiques de la modélisation et de la simulation,* edited by Franck Varenne and Marc Siberstein, 145–70. Paris: Editions Materiologiques.

Pittendrigh, Collin. 1954. "On Temperature Independence in the Clock System Controlling Emergence Time in *Drosophila*." *Proceedings of the National Academy of Sciences of the United States of America* 40: 1018–29.

Sprinzak, David, and Michael B. Elowitz. 2005. "Reconstruction of Genetic Circuits." *Nature* 438: 443–38.

Spudich, J. L., and D. E. Koshland. 1976. "Non-Genetic Individuality: Chance in the Single Cell." *Nature* 262: 467–71.

Stricker, Jesse, Scott Cookson, Matthew R. Bennet, William H. Mather, Lev S. Tsimring, and Jeff Hasty. 2008. "A Fast, Robust and Tunable Synthetic Gene Oscillator." *Nature* 456: 516–19.

Swain, Peter S., Michael B. Elowitz, and Eric D. Siggia. 2002. "Intrinsic and Extrinsic Contributions to Stochasticity in Gene Expression." *Proceedings of the National Academy of Sciences of the United States of America* 99: 12795–800.

Swoyer, Chris. 1991. "Structural Representation and Surrogative Reasoning." *Synthese* 87: 449–508.

Waters, C. Kenneth. 2012. "Experimental Modeling as a Form of Theoretical Modeling." Paper presented at the 23rd Annual Meeting of the Philosophy of Science Association, San Diego, Calif., November 15–17, 2012.

Weber, Marcel. 2012. "Experimental Modeling: Exemplification and Representation as Theorizing Strategies." Paper presented at the 23rd Annual Meeting of the Philosophy of Science Association, San Diego, Calif., November 15–17, 2012.

Winsberg, Eric. 2003. "Simulated Experiments: Methodology for a Virtual World." *Philosophy of Science* 70: 105–25.

Winsberg, Eric. 2009. "A Tale of Two Methods." *Synthese* 169: 575–92.

5

WILL YOUR POLICY WORK?
Experiments versus Models

NANCY D. CARTWRIGHT

WHAT GETS LABELED "GOOD EVIDENCE"?

Suppose you want to improve the mathematics achievements of students in your high school. Here in the United States you are urged to adopt only programs that are well evidenced to work. To find those you can go to the Institute of Education Science's What Works Clearinghouse (WWC), set up by the U.S. Department of Education. The WWC website has a section called "intervention reports," where an intervention report, by its own description, is "a summary of findings of the highest-quality research on a given program, practice, or policy in education."[1] Type in "mathematics, high school," and you find fourteen results—or at least I did when I checked just before Valentine's day in 2013. Of the fourteen, two were for middle, not high school; one was for pre-K, one for learning disabilities, one about reading for third graders, one for dropout prevention, one where math was one aspect of the program but the study on the math aspect did not meet WWC standards and so was not counted, and one on math where sixteen studies were reviewed but none met WWC standards so the verdict is "unable to draw a conclusion."

That leaves six. Each of these six programs has a scorecard that provides a full report and a report summary of "effectiveness," plus a scorecard with an "Improvement Index (percentile gain for the average student)," an "Effectiveness Rating," and a rating of the "Extent of Evidence." The WWC scorecard for Core-Plus Mathematics, for instance, states that the program "was found to have potentially positive effects on mathematics achievement for high school students" and gives it an "Effectiveness Rating" of "+" (on a scale from −− to ++).[2] It also gives Core-Plus Mathematics "15" on the "Improvement Index," though the extent of the evidence is judged "small." Of the other programs, three have essentially a zero effectiveness rating (two on small and one on medium-large evidence), and the other two have positive effectiveness ratings but are rated "small" on evidence. That makes three altogether that are judged to have good evidence—though small in extent—of being positively effective in study settings. Not much to choose from!

What then is good evidence? What does WWC count as "highest quality" research? They count evidence from experiments of a very special kind—randomized controlled trials (RCTs). They tell us, "Currently, only well-designed and well-implemented randomized controlled trials (RCTs) are considered strong evidence" (What Works Clearinghouse 2011, 11). Randomized controlled trials do have special virtues: in the ideal, they can clinch causal conclusions about the study population. If there is a higher average effect in the treatment group than in the control group, we are assured that the treatment helped at least someone in the study population. We can be assured of this supposing that the methods prescribed for proper conduct of an RCT, such as masking and randomization, have achieved their goals, and that all factors other than the treatment (and its downstream effects) that affect the outcome have the same distribution in the treatment and control groups (in expectation), and supposing that our methods for statistical inference have yielded the true averages. In practice, it is very difficult to do just what is required for a good RCT, which is why so few studies pass the "good evidence" test employed by WWC and hosts of similar clearing houses for policies ranging from education to public health to criminal justice.

NECESSARY MAYBE BUT NOT SUFFICIENT

Notice that the advice to choose only policies that have been shown effective in studies of the highest research quality gives only a necessary condition: do not use it unless it meets these standards. It does not give a positive direction: do use it if it does meet these standards. That is because, as is well known, there can be important, relevant differences between the study population and yours. The California class-size reduction program is a notorious example.

A good RCT in Tennessee showed that reducing class sizes improved a number of educational outcomes in that state. But reducing class size in California generally failed to improve educational outcomes, and in some cases even made them worse. What would have helped California to make less optimistic predictions about the success of their program? There are some who answer "more experiments," and especially experiments in other places than Tennessee, maybe even in California itself. This, I claim, is a mistake. What California needed was not an experiment but a model, a causal model. Let me explain.

Suppose there were a few further RCTs conducted on reducing class size, maybe in a state with a large non–English-speaking school population as in California, and one with a high technology segment as in California, and one on the Pacific Rim like California. These are the kinds of things that the casual advice usually available might lead you to think are relevant to predicting whether the policy will succeed in California. But there is no call to be casual. We can know, at least in the abstract, just what it takes for the experiments to be relevant. If the settings for all these studies were *felicitous* for the causal pathway class-size reduction → better educational outcomes— where I can tell you just what constitutes felicity[3]—then in all these RCTs class-size reduction would have proven effective. If the setting for any of them was infelicitous, the result might have been negligible or even negative.

At least a negative result would have sounded a warning bell; perhaps if all these had negative results and only Tennessee's were positive, small classes would never make it through a WWC review. But this would be a shame because there are good reasons to think that reducing class size can have a big positive effect if done in the right circumstances—that is, in circumstances that are, in my technical sense, "felicitous."

Nor should you think that results like this would be surprising. Mixed results in good RCTs for the same social program across different settings are not at all unusual. Consider, for instance, the conflicting results produced by two of the RCTs included in a Cochrane Collaboration meta-analysis on the effects of introducing safety regulations on fatal work-related injuries. According to one RCT, introducing safety regulations led to a decrease in such injuries, with a standardized mean difference (SMD) between treatment and control of −1.4. However, according to another RCT, introducing safety regulations led to an increase in fatal work-related injuries (SMD = +2.39) (van der Molen et al. 2012, 28). Or, to take another example, consider a Cochrane Collaboration meta-analysis on the effect of exercise on depression. According to one of the RCTs included in this meta-analysis, exercise has an important negative effect on depression symptoms (SMD = +1.51); according to another included RCT, exercise has an important positive effect (SMD = −1.16) on depression symptoms (Rimer et al. 2012, 87).

I should add a cautionary note though. When we do see mixed results, they are often not really from comparable experiments. The exact protocol for the treatment may vary; the procedures for measuring outcomes and sometimes even the outcomes themselves also may vary. Also, often the nature of the control varies. For social programs it is usual for the control group to receive the standard available alternative program. As my British associates are fond of pointing out, this often means that programs that test effective in the United States turn out to test ineffective in the United Kingdom. The reason for the difference in study results is that the social services otherwise available are generally much better in the United Kingdom than in the United States.

Getting back to California and Tennessee, let us consider what went wrong in California that had gone right in Tennessee. Two factors stand out in the later analysis of why the California program did not achieve its intended results—availability of space and of good teachers. California implemented the statewide reduction over a very short period of time. The legislation was passed in the spring of 1996, and class sizes were widely reduced already by the fall of the same year. Smaller classes mean more classes, and more classes require more teachers and more classrooms. Tennessee had enough good teachers and enough space to manage. California did not. California had to find an additional 18,400 teachers for the fall. As a result, many poorly qualified teachers were hired, and many of the least qualified ended up in the most disadvantaged schools. Classes were held in inappropriate places, and space was also taken from other activities that promote learning indirectly. These are just the kinds of factors I would expect to see in a good causal model of the California program. And we need these models because, as a Wisconsin Policy Research Institute report (Hruz 2000, 1) underlines, "One critical lesson that can be drawn from both the national research on class size policies and the results of Wisconsin's own SAGE program is that smaller classes do not always provide identifiable achievement benefits."

PURPOSE-BUILT CAUSAL MODELS

So what is a causal model? A lot of things can go under this label. Often what we call *causal models* in the social sciences describe facts about *general* causal relations, what we sometimes call *causal principles* or *causal laws*. In the social sciences, statistics are generally the central immediate empirical input for these models. In economics, econometric models come in the form of sets of equations, which are often regression equations with parameters estimated from the data. These can turn into causal models when appropriate assumptions can be added about some of the causal relations modeled.[4] Causal path diagrams used to be popular in sociology. Nowadays various groups of philosophers, working, for instance, with Clark Glymour, Peter Spirtes, or Judea Pearl, will be familiar with modeling causal principles with directed acyclic graphs, which they use Bayes-nets methods to construct from statistical data, previous causal knowledge, and a handful of crucial assumptions about the relations between causes and probability.

Generic causal models represent facts about general causal principles. But how general? If done well, they will at least represent the causal principles that held at the time in the population from which the data were drawn and of which any additional assumptions necessarily, by construction, are true. What about more widely? Here we often encounter the same misleading advertising as with WWC, where "what works" surely implies a fair degree of generality but what is vetted is whether a policy has worked somewhere. Similarly with a great many generic models found in the social sciences, the manner in which they are presented and the language used often suggest they should apply outside the population from which the data are drawn—not universally perhaps, but at least to some wide range of cases that meet some happy set of criteria. The trouble is that we are rarely told what those criteria are. It is even more rare to find justification for why any criteria proffered are the ones that pick out circumstances in which the model should hold.

There is a straightforward answer to where the model will hold: it will hold in any population in which (1) the same probabilistic relations hold that were inferred from the original data and are necessary to construct the model; (2) any facts about causal relations that are necessary for constructing the model hold; and (3) all additional assumptions necessary hold as well. For econometric models, the latter generally include assumptions about functional form; for getting directed acyclic graphs by Bayes-nets methods, it includes any of the axioms (like the causal Markov condition) needed to prove that the causal conclusions represented in the model can be inferred from the input information.

This answer is highly abstract, perhaps too abstract to be of much practical use. But even this straightforward answer is never articulated, and I have seldom seen it implicit in the discussion when models are presented. This is a serious failing. These are necessary and essentially sufficient condi-

tions for the facts in a model constructed from data from one setting to obtain in another. (I say "essentially" because the facts from the original
setting *could* otherwise hold elsewhere, but that would be by accident, having nothing to do with what justifies the model for the original setting in
the first place.) Because they are necessary and sufficient conditions, they
cannot be ignored in any attempt to defend the range of the model's applicability: any argument that the model applies across a given range must be
sufficient to show that these three conditions are met.

This is enough about generic causal models, however; I mention them at
this point only to dismiss them. When I say that policy users need a causal
model, not an experiment, it is not generic causal models I intend, but rather
what I shall call purpose-built single-case causal models. The important difference is that the generic causal model for a given situation presents facts
about what causal relations *can* obtain there. The policy maker for the situation needs information about what *will* (and will not) happen there. This is
what a purpose-built single-case causal model is for.

"Policy maker" is not a homogenous category. It includes the California
legislature considering whether to mandate smaller classes in the fall, high
school principals considering whether reducing classes would be likely to
produce improvement in their schools, and parents considering whether it
will help their child. The first two may be able to make do with averages,
though for populations very different in size. The parents want to know
about the effects on their child.

Randomized controlled trials only give averages, so an RCT result will
not tell the parents what they want to know. That is a well-known problem,
and it is an important problem—but it is not mine here. When I talk about
models that are single case and purpose built I am not pointing to the difference between a model for the individual parent and a model for the legislature. Both need a model that is purpose built for their case, which is a
specific *single* case: a policy implemented. Single-case purpose-built causal
models can come in a variety of different forms that can do different jobs.
Let me illustrate with a few forms to provide an idea of what they are.

Figure 5.1 shows a simple model for the California class-size reduction
program that uses what epidemiologists call "causal pies" but I call "causal
pancakes." Each slice of Figure 5.1 represents a factor that must be present if
the salient factor (in our case the policy of reducing class sizes) is to achieve
the desired outcome. Jeremy Hardie and I call these other necessary factors
support factors for the policy to produce the outcome.[5] We use the label

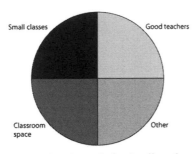

Figure 5.1. Causal pancake for the effect of class size reduction on learning outcomes.

"pancakes" rather than "pies" because you may be thinking that if you take away one slice of a pumpkin pie you still have quite a lot of pie left—which is just the wrong idea. The point is that these are necessary factors that must be in place if the policy is to produce the desired effect level. The analogy is with the ingredients of a pancake: you need flour, milk, eggs, and baking powder. If you take away one of these four ingredients, you do not get 3/4 of a pancake—you get no pancake at all. It works similarly with small classes: if you take away the good teachers or the adequate classroom space, you do not get 2/3 of the improvements expected—you get no desired improvement at all.

This model for the California case does warn of the need for enough teachers and enough classroom space, which were likely to be missing given the envisaged methods and timing of implementation. But it does not warn about another problem that occurred: classrooms were taken away from other activities that promote learning. As shown in Figure 5.2, we can expand the model to represent these problems by adding other pancakes. Pancakes B and C show other sets of factors that in the setting in California in 1996 would help improve the targeted educational outcomes but that the program implemented as envisaged would have a negative impact on. You can see that appropriate space is required for all of them to do the job expected of them, which was just what California was not likely to have once it reduced class sizes statewide over a short period of time.

I said that purpose-built causal models can come in a variety of forms. A causal loop model, which is more sophisticated, shows more details about what is needed and the roles, both positive and negative, that the various fac-

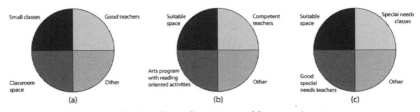

Figure 5.2. Causal pancakes for the effects of various sets of factors on learning outcomes.

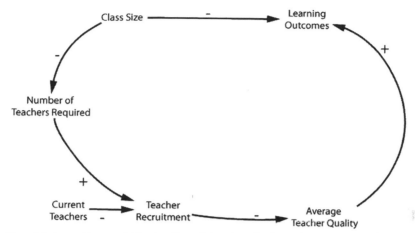

Figure 5.3. A causal loop model for the effect of class size reduction on learning outcomes (by David Lane). Note that + means a positive influence and − is a negative.

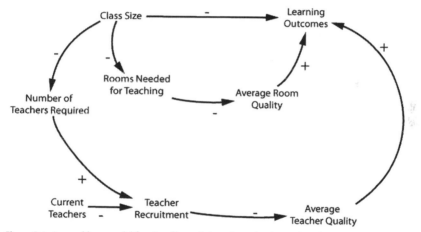

Figure 5.4. A causal loop model for the effect of class size reduction on learning outcomes that incorporates room quality. (David Lane)

tors play.[6] They are called loop diagrams not because they use curved arrows, which can make processes more visible, but more especially because they allow the depiction of feedback loops. In Figure 5.3 we see how the problem with teachers unfolds. Then, as shown in Figure 5.4, we can add a pathway showing how learning outcomes will be negatively affected by the reduction in room quality induced by the policy. Last, we show how the policy produces negative outcomes by removing rooms from other learning-supporting activities, as seen in Figure 5.5.

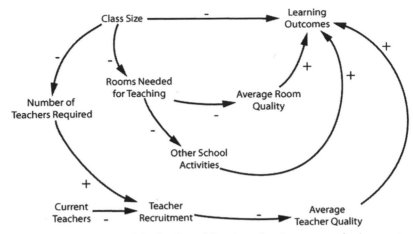

Figure 5.5. A causal loop model for the effect of class size on learning outcomes that incorporates room quality and other school activities. (David Lane)

Notice that these different methods of representation have different advantages and disadvantages. Causal loop diagrams show more steps and hence give more information. For instance, the loop diagrams warn us that it is not enough to have sufficient classrooms in California after implementation—the classrooms also must be of sufficient quality. And the role of teacher recruitment in the process and how that is to be accomplished is brought to the fore. Perhaps with this warning California could have invested far more in recruiting teachers nationwide.

On the other hand, you cannot tell from the loop diagram that, as we hypothesized earlier, neither a sufficient number of good quality classrooms by itself along with smaller classes, nor a sufficient number of good quality teachers along with smaller classes will do; rather, all three are necessary together. This kind of information is awkward to express in a loop diagram, but it is just what pancake models are designed to do.

There is one further thing that loop diagrams can do that cakes are not immediately equipped for: to represent the side effects of the implemented policy and the processes by which these are produced. That is because an array of pancakes is supposed to represent sets of jointly necessary factors, each set sufficient for the production of the *same* effect. But the side effects produced by a policy are clearly relevant to the deliberations about adopting that policy if they embody significant benefits or harms, even if those side effects are not directly relevant to the narrow effectiveness question of "Will the policy as implemented produce its intended effects?"

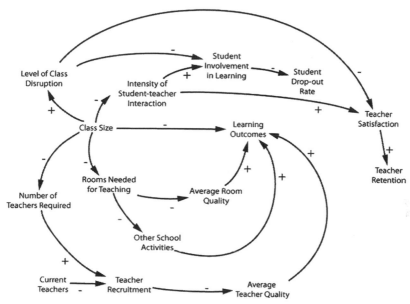

Figure 5.6. A causal loop model for the effect of class size reduction on learning outcomes which shows beneficial side effects. (David Lane)

Lower dropout rates and higher teacher retention rates are two beneficial side effects. There was good reason to think these would apply to California, despite the failure of smaller classes to produce significant improvements in the targeted educational outcomes envisaged in the purpose-built models we have examined.[7] Figure 5.6 is a diagram to represent the production of these beneficial side effects.

A third possibility is that, with enough detailed information and background theory to build from, one might offer a mathematical model to represent some of the causal pathways and to provide numerical estimates. I can offer a sample of what such a model might have looked like for the California case. In the first year 18,400 classes were added, an increase of 28 percent, to get class size down to twenty students or under. These are just the numbers that should appear in the model, and beyond that I do not mean to suggest that the functions offered in the model are correct—to determine that would take a more serious modeling exercise. I simply filled in some possible functional forms to provide a concrete idea what a purpose-built mathematical model might look like.

A Hypothetical Mathematical Model for California's Class Size Reduction Program

Let B = number of classes before program start (= 65,714); r = fraction by which class size is reduced (= 0.78); T = number of new good teachers available at the time of program start; LR = number of rooms devoted to supportive learning activities before program start = number of new rooms available for ordinary classes at time of program start; TQ = teacher quality after program implementation; RQ = room quality after implementation; LA = number of supportive learning activities after program implementation; and I = improvement in targeted learning outcomes. B, r, T, and LR are to be filled in with actual numerical values.

Supposing there was and, as the legislation requires, will continue to be 1 teacher and 1 classroom per class in California at the time, then the number of new rooms needed = the number of new teachers needed = $B(1/r - 1) = 65,714(1/0.78 - 1) \approx 18,400$.

$$TQ = \alpha\, e^{T - B(1/r - 1)} \quad \text{if} \quad T \leq B(1/r - 1)$$
$$RQ = \beta\, e^{LR - B(1/r - 1)} \quad \text{if} \quad LR \leq B(1/r - 1)$$
$$LA = 0 \quad \text{if} \quad LR \leq B(1/r - 1)$$
$$I = \gamma(TQ \times RQ)/r + \delta LA = \varepsilon + 0, \varepsilon \text{ small}$$

The examples I hope make it clear what I mean by a purpose-built single-case model. A *purpose-built single-case causal model* for a given cause-cum-implementation C_i, a given effect E, and a given situation S encodes information about factors (or disjunctions of factors) that must occur if C_i is to contribute to E in S, or factors that C_i will affect in S. As we have seen, purpose-built causal models can come in a variety of forms, and they can convey different items and different amounts of information. The good ones convey just the information you need so that you can avoid making mistaken predictions. With the examples we have looked at in mind, I suggest that a purpose-built causal model is *most useful* when it tells you:

1. The factors (or disjunctions of factors) that are necessary if C_i is to contribute to E in S but that might not occur there.
2. The effects of C_i on other factors that would contribute (either positively or negatively) to E were C_i absent.

3. Any significant deleterious or beneficial factors to which C_i (in conjunction with other factors present in S after implementation) will contribute in S.

And if it is possible to be quantitative:

4. The size of C_i's overall contribution to E in S both from its direct influence on E and from its indirect influence on E, via its influence on other factors that would contribute to E in its absence.
5. The size of C_i's contributions to other significant deleterious or beneficial factors that will occur in S after implementation.

Note that in item 1 when I say "might" I do not mean "might so far as we know" though this is always a central issue. Rather I mean a *might* that is rooted in the structure of the situation—a real possibility given the circumstances. In California, the lack of classroom space and the shortage of teachers once the program was in place were facts very likely to materialize given that the program was to be rolled out statewide over such a short time, unless something very special was done to avert them. These are the kinds of facts that we hope the modeler will dig out while preparing to produce the model that we must rely on for predicting the outcomes of our proposed policy.

Readers will have noticed that my purpose-built single-case models look very different from what philosophers of science—me or James Woodward or Judea Pearl, for example—usually talk about under the heading "causal models," which look like this:

GCM:

$$X_1 = u_1$$

$$X_2 = f_{21}(B_{21}, X_1) + u_2$$

$$\cdots$$

$$X_n = \sum_{i=1}^{n-1} f_{ni}(B_{ni}, X_i) + u_n$$

That is because the usual topic is generic causal models that, as I described earlier, are models that give information about the general causal principles that dictate what causal processes can happen in a situation, whereas a single-case purpose-built model tells about what does happen.

Let us look at the model form GCM to note some of the differences. The first thing to note is that the equations in a model of form GCM are always *ceteris paribus*. They hold only so long as nothing untoward happens. Modelers often, I think mistakenly, try to represent this by adding an additional variable at the end of each equation to stand for "unknown factors."[8] *B* rarely represents a single factor. Rather, B_{ji} represents all the separate clusters that in conjunction with x_i are sufficient to produce a contribution to x_j. That is a lot to find out about. And no matter how thorough we are, whatever we write down will be correct only ceteris paribus. With the possible exception of the basic causal laws of physics, there are always features—probably infinities of them—that can disrupt any causal principle we have written down.

The modeler building specifically for our particular case has it easier. Among the necessary support factors for x_i to contribute to x_j—that is, among all those factors that are represented in B_{ji}—to be maximally useful, the model need only diagram factors that, in the actual facts of the situation, might not be there. As to factors represented in the other terms of the sum for x_j, it need only include those that the implementation of x_i might impact on. And it need take note of factors stashed away in the ceteris paribus clause only if they might actually obtain.[9]

Of course this is no gain if the only way to know what these factors are is to first establish a generic model like GCM for situations like ours; then consult this generic model to extract all the factors described there; then look to see whether any of these might or might not obtain. But that is not necessary. We are very often in a good position to see that a factor will be necessary in situ without having a general theory about it. The same is the case with possible interferences. I do not need to know much about how my cell phone works nor all the infinitude of things that could make it stop doing so in order to understand that I should not let it drop out of my pocket into the waterfall. Nor do I have to know the correct functional form of the relation between teacher quality, classroom quality, class size, and educational outcomes to suggest correctly, as my sample mathematical model does, that outcomes depend on a function that (within the ranges occurring in mid-1990s California) increases as teacher and room quality increase, and decreases as class size increases.

How do we—or better, some subject-specific expert modeler—come to know these things? What can give us reasonable confidence that we, or they, have likely gotten it right, or right enough? I do not know. I do know that we have, across a great variety of subject areas, expert modelers who are good at the job, albeit far from perfect. And all of us do this kind of causal modeling successfully in our daily lives all the time: from calculating how long it will take to drive to the airport at a given time to determining whether we can empty the garbage in our pajamas without being observed. We do not, though, have so far as I know any reasonable philosophic accounts of how this is done, either in general or in specific cases.[10] This is, I urge, one of the major tasks philosophers of science should be tackling now.

But wanting a theory should not make us deny the fact. We are often able to build very useful models of concrete situations that can provide relatively reliable predictions about the results of our policies. Obviously in doing so we rely not just on knowledge of the concrete situation but heavily on far more general knowledge as well. We do not need generic models of situations of the same general kind first.

COMPARING: EXPERIMENTS VERSUS MODELS

Let us return to the major issue, models versus experiments. When I talk about the kinds of facts that "we *must* rely on" for good prediction, that is just what I mean. What we see in what I have described as a maximally useful purpose-built model is just the information that Nature will use in producing effects from what we propose to do in implementing our policy. Whether we want to build this model or not and whether we are willing or able to spend the time, the money, and the effort to do so, what is represented in the model is what matters to whether we get our outcomes. It is what Nature will do in our situation, whether we figure it out or not. I take it then that responsible policy making requires us to estimate as best we can what is in this model, within the bounds of our capabilities and constraints.

That is the argument for modeling. What is the argument for experiments, in particular for RCTs, which are the study design of choice in the community who rightly urges we should use evidence to inform predictions about policy outcomes? There is one good argument for experiments, I think, and one bad.

The good argument has to do with certainty. Results of good experiments can be more certain than those of good modelers. In the ideal, recall that an RCT can clinch its conclusion. What is nice about the RCT is that what counts as doing it well is laid out in detailed manuals, which can be consulted by outsiders at least to some reasonable extent. This is not the case with modeling, where there is no recipe. Modeling requires training, detailed knowledge, and good judgment, all of which are hard to police. A good modeler should of course be able to provide reasons for what appears in the model, but there will be no rule book that tells us whether these reasons support the judgments that the modeler makes.

So in general, insofar as it makes sense to talk like this, the results of a well-conducted RCT can probably be awarded more credibility than the models of even a modeler who is very good at building just this kind of model for just this kind of situation.

However, there is a sting in the tail of this good argument in favor of the experiment over the model. The good experiment provides a high degree of certainty but about the wrong conclusion; the model provides less certainty but about the right conclusion. The experiment tells about whether the policy produced the outcome for some individuals in the study situation. And, of course, it can be of real use when building a model to know that the policy worked somewhere, especially if we can come to have more understanding of how it worked where it did. After all, as Curtis Meinert, who is an expert on clinical trial methodology, points out, "There is no point in worrying about whether a treatment works the same or differently in men and women until it has been shown to work in someone" (Meinert 1995, 795). So the experiment can provide significant suggestions about what policies to consider, but it is the model that tells us whether the policy will produce the outcome for us.

There is one experiment that could tell us about our situation—if only we could do this experiment, at least so long as we are only interested in averages in our situation (as the state of California might be, but an individual principal or an individual parent might not). That is to do an experiment on a *representative sample* from *our population*—that is, the population who will be subject to the policy. That works by definition because a representative sample is, by definition, one with the same mean as the underlying population. The favored way to do this is by random sampling from the target, although we need to take this with a grain of salt because we know that genuine random sampling will often not produce a representative sample. This kind of experiment is generally very difficult to set up for practical reasons. And random sampling from the target is usually very difficult to arrange for a study, for a variety of reasons. For example, studies must be performed with individuals the study can reach and monitor; for ethical reasons, the individuals who are enrolled must consent to be in the study; often subgroups of the population who in the end will be subject to the policy are excluded from the study for both ethical reasons and in order to see the "pure" effects of the program (such as when the elderly are excluded from medical studies because the constraints of the study may be harmful, or individuals taking other drugs may be excluded so that the "pure" effects of the drug unadulterated by interaction with other drugs may be seen).

Even if there are no constraints and a random sample is drawn, the mere fact that it is a sample can affect the outcome. So the results from the random sample will not match those that would follow were the entire target population to be subject to the policy. If only a portion of schools in the state have their class size reduced, there may be enough teachers to staff the increased number of classes; or if only random districts are involved, the parents who can afford to move may do so, to take advantage of the smaller classes in the "treated" districts.[11] Time also has its effects. Time passes between the experiment and the policy implementation, and time brings changes—sometimes significant changes—to what matters to whether the policy will produce the desired outcomes.

If the experimental population is not drawn as a random sample from the target, then predicting that our situation will have similar results as in the experiment is dicey indeed. The RCT estimates average results—the average value of the outcome in the treatment and in the control group. Everyone admits that it deals with averages. The problem is what the averages are over. Recall those support factors that are necessary for the policy to produce the outcome? It is easy to show that these are *just* what the RCT averages over.[12] It averages over all the different combinations of values these support factors take in the study population. The whole point of the RCT is to finesse the fact that we do not know what these factors are but we find out whether the policy can produce the outcome sometimes. However, it is no good trying to get along with our ignorance of these factors—their distribution will fix the averages we see. And, as we have seen represented in our models, if the support factors are all missing in our situation, the policy will not produce the outcome at all for us.

That was the good argument for RCTs. Now here is the bad one. Randomized controlled trials can give us effect sizes, and we need effect sizes for cost-benefit analyses, and we need cost-benefit analyses for responsible comparison of policies. This is how Michael Rawlins, head of the U.K. National Institute for Health and Clinical Excellence, defends the need for them.[13] Or in the United States, let us look at an extended example, the lengthy Washington State Institute for Public Policy document on how the institute proposes to conduct cost-benefit analyses of policies the state considers undertaking. They admit, at step 3 (out of 3), that "almost every step involves at least some level of uncertainty" (Aos et al. 2011, 5). They do try to control for this and to estimate how bad it is. They tell us the following about their procedures:

- "To assess risk, we perform a 'Monte Carlo simulation' technique in which we vary the key factors in our calculations" (7).
- "Each simulation run draws randomly from estimated distributions around the following list of inputs" (122).

Among these inputs is our topic at the moment, effect sizes:

- "Program Effect Sizes . . . The model is driven by the estimated effects of programs and policies on . . . outcomes. We estimate these effect sizes meta-analytically, and that process produces a random effects standard error around the effect size" (122).

Earlier they told us that they do not require random assignment for studies in the meta-analysis, but they do use only "outcome evaluation[s] that had a comparison group" (8). A meta-analysis is a statistical analysis that combines different studies to produce a kind of amalgam of their results.

What this procedure does, in effect, is to suppose that the effect size for the proposed policy would be that given by a random draw from the distribution of effect sizes that come out of their meta-analysis of the good comparative studies that have been done. What makes that a good idea? Recall that the effect size is the difference between the average value of the outcomes in treatment and control groups. And recall that this average is determined by the distribution of what I have called "support factors" for the policy to produce the targeted outcome. The meta-analysis studies a large population—the combined population from all the studies involved—so it has a better chance of approaching the true averages for that population. But that is irrelevant to the issue of whether the distribution of support factors in your population is like that from the meta-analysis. Or is yours like the—possibly very different—distribution of the support factors in some particular study that participates in the meta-analysis, which could be very different from that of the meta-analysis? Or is yours different from any of these? Without further information, what justifies betting on one of these over another?

The bottom line is that, yes, RCTs do indeed give effect sizes, and you would like an effect size for your case in order to do a cost-benefit analysis. But effect size depends on the distribution of support factors, and this is just the kind of thing that is very likely to differ from place to place, circumstance to circumstance, time to time, and population to population. You cannot use the need for an effect size here as an argument for doing an RCT somewhere else, not without a great deal more knowledge of the facts about here and the somewhere else. And somewhere else is inevitably where the studies will be done.

I BEGAN BY POINTING out that good experimental evidence about a policy's effectiveness is hard to come by, and it can only tell you what outcome to expect from the same policy in your setting if the relations between the experimental setting and yours are felicitous. Thus, for predicting the outcomes of proposed policies, experiments can provide highly credible information—that may or may not be relevant. Whether they give much or little information depends on the distribution of support factors you have in your situation. Experiments are silent about that: they tell you neither what these support factors are nor what you have. For reliable prediction, a single-case causal model is essential. Whether or not you are able to come by a good one, it is the information in a model of this kind that will determine what outcomes you will obtain.

Here I have described what a single-case purpose-built causal model is, explained how single-case purpose-built models differ from the generic models that are so much talked of by philosophers nowadays, shown how purpose-built models can differ in form and supply different kinds of information relevant to what outcome you will obtain, and laid out just what information a model maximally useful for predicting your outcome should contain. My final conclusion can be summed up in a few words: for policy deliberation, experiments may be of use, but models are essential.

NOTES

I would like to thank Alex Marcellesi and Joyce Havstad for research assistance as well as the Templeton project God's Order, Man's Order, and the Order of Nature and the AHRC project Choices of Evidence: Tacit Philosophical Assumptions in Debates on Evidence-Based Practice in Children's Welfare Services for support for research for this essay. I would also like to thank the University of Dallas Science Conversations for the chance to discuss it.

1. "Glossary," What Works Clearinghouse, http://ies.ed.gov/ncee/wwc/glossary.aspx.

2. "Intervention > Evidence Report, Core-Plus Mathematics" IES What Works Clearinghouse, http://ies.ed.gov/ncee/wwc/interventionreport.aspx?sid=118.

3. See Cartwright and Hardie (2012) for sufficient conditions for a program to make the same contribution in new settings as it does in a particular study setting (note that the term "felicitous" does not appear there).

4. For examples of what these additional requirements might be, see Cartwright (2004) or Reiss (2005).

5. See Cartwright and Hardie (2012), chapter II.A.

6. These diagrams have been constructed by David Lane of Henley Business School from the discussion in Cartwright and Hardie (2012), §I.A.1.2.

7. I haven't found evidence about whether these did occur in California. There is good evidence that these have been effects of smaller classes elsewhere. Of course the lack of good teachers and appropriate space could have undermined these processes as well . . . And that could be readily represented in a causal loop diagram.

8. It is a mistake because these kinds of causal principles depend on, and arise out of, more underlying structures, which it is very misleading to represent with variables of the kind appearing in the equations. See Cartwright (2007) and Cartwright and Pemberton (2013).

9. I first realized this advantage of the single-case model over generic models for systems of the same kind by reading Gabrielle Contessa's account of dispositions (2013).

10. Ray Pawson (2013) makes an excellent offering in this direction.

11. Thanks to Justin Fisher for reminding me to add these possibilities.

12. See Cartwright and Hardie (2012) or Cartwright (2012) for proofs and references.

13. In a talk entitled "Talking Therapies: What Counts as Credible Evidence?" delivered during the "Psychological Therapies in the NHS" conference at Savoy Place in London in November 2011.

REFERENCES

Aos, Steve, Stephanie Lee, Elizabeth Drake, Annie Pennucci, Tali Klima, Marna Miller, Laurie Anderson, Jim Mayfield, and Mason Burley. 2011. *Return on*

Investment: Evidence-Based Options to Improve Statewide Outcomes. Technical Appendix II: Methods and User-Manual. Olympia: Washington State Institute for Public Policy.

Cartwright, Nancy D. 2004. "Two Theorems on Invariance and Causality." *Philosophy of Science* 70: 203–24.

Cartwright, Nancy D. 2007. *Hunting Causes and Using Them.* Cambridge: Cambridge University Press.

Cartwright, Nancy D. 2012. "Presidential Address: Will This Policy Work for You? Predicting Effectiveness Better: How Philosophy Helps." *Philosophy of Science* 79: 973–89.

Cartwright, Nancy D., and Jeremy Hardie. 2012. *Evidence-Based Policy: A Practical Guide to Doing It Better.* New York: Oxford University Press.

Cartwright, Nancy D., and J. Pemberton. 2013. "Causal Powers: Without Them, What Would Causal Laws Do?" In *Powers and Capacities in Philosophy: The New Aristotelianism,* edited by Ruth Groff and John Greco, 93–112. New York: Routledge.

Contessa, Gabriele. 2013. "Dispositions and Interferences." *Philosophical Studies* 165: 401–19.

Hruz, Thomas. 2000. *The Costs and Benefits of Smaller Classes in Wisconsin: A Further Evaluation of the SAGE Program.* Wisconsin Policy Institute Report, vol. 13, issue 6. Thiensville, WI: Wisconsin Policy Research Institute.

Meinert, Curtis L. 1995. "The Inclusion of Women in Clinical Trials." *Science* 269: 795–96.

Pawson, Ray. 2013. *The Science of Evaluation: A Realist Manifesto.* Los Angeles: SAGE.

Reiss, Julian. 2005. "Causal Instrumental Variables and Interventions." *Philosophy of Science* 72: 964–76.

Rimer, Jane, Kerry Dwan, Debbie A. Lawlor, Carolyn A. Greig, Marion McMurdo, Wendy Morley, and Gillian E. Mead. 2012. "Exercise for Depression." *Cochrane Database of Systematic Reviews,* no. 7: CD004366.

van der Molen, Henk F., Marika M. Lehtola, Jorma Lappalainen, Peter L. T. Hoonakker, Hongwei Hsiao, Roger Haslam, Andrew R. Hale, Monique H. W. Frings-Dresen, and Jos H. Verbeek. 2012. "Interventions to Prevent Injuries in Construction Workers." *Cochrane Database of Systematic Reviews,* no. 12: CD006251.

What Works Clearinghouse. 2011. *Procedures and Standards Handbook Version 2.1.* Washington, D.C.: U.S. Department of Education, Institute of Education Sciences.

6

CAUSAL CONTENT AND GLOBAL LAWS
Grounding Modality in Experimental Practice

JENANN ISMAEL

W HEN PHILOSOPHERS APPROACH SCIENCE, there is a strong tendency to focus on the products of science rather than the practice. What follows is a case study in how that focus has distorted the philosophical discussion of science and laws, and how an emphasis on the experimental side of modeling resolves them. The focus on the products of science leads to a fixation on global models and a tendency to treat them representationally. An emphasis on the experimental side of modeling, by contrast, directs our gaze toward models of open subsystems and gives us the tools for a more pragmatic approach to modal content.

There was a time when science was thought of almost exclusively in causal terms. The mandate of science was thought to be the investigation of the causal structure of the world. Things changed with the mathematicization of science and the triumph of Newtonian theory. Newton's theory provided dynamical laws expressed in the form of differential equations that could be used to compute the state of the world at one time as a function of its state at another. Theoretical developments since Newton have seen important departures. Quantum mechanics introduced indeterminism. We are still far from a fundamental theory, but in the philosophical literature, Newtonian theory still serves in many circles as a paradigm for what a fundamental theory should look like: it should be global in scope, and the fundamental laws should take the form of what we call equations of motion, which is to say they tell us how the state of the world evolves from one moment to the next.

The eclipsing of causal notions in physics happened almost unnoticed until Russell's justly famous paper of 1913 in which he detailed the differences between the two notions (Russell 1913/1953). Russell himself thought that the differences were so great, and that the causal notions carried so many ill-fitting associations, that they should be eliminated from exact science. That early position was rebutted in a paper by Nancy Cartwright (1979) in which she argued that causal knowledge is indispensable in practical reasoning. And so began the long struggle to understand how causal ideas enter into scientific description.[1]

REDUCTIVE PROJECTS

Reductive projects dominate the rather large body of post-Russellian discussion of causation, especially among philosophers of physics. The thought is that what happened with cause is what happened with so many other notions once thought to be basic to our understanding of nature. Once the physics has progressed so that cause no longer makes an appearance at the fundamental level, there is room for an illuminating reduction, so we should be looking to reduce causal structure to physical laws. And the presumption was that these laws should take the form of global laws of temporal evolution, modeled on the Newtonian law of gravitation. These were the kinds of laws that Russell regarded as the most basic modal generalizations in physics. He wrote,

> In the motions of mutually gravitating bodies, there is nothing that can be called a cause, and nothing that can be called an effect; there is merely a formula. Certain differential equations can be found, which hold at every instant for every particle of the system, and which, given the configuration and velocities at one instant, or the configurations at two instants, render the configuration at any other earlier or later instant theoretically calculable. That is to say, the configuration at any instant is a function of that instant and the configurations at two given instants. (Russell 1913/1953, 14)

Let us call the assumption that the most basic laws take the form of global laws of temporal evolution—that is, laws that give the state of the universe at one time as a function of its state at another—*globalism*. Most philosophers have followed Russell in presuming globalism. I say "presuming" because globalism is almost without exception assumed without argument, and it plays a role shaping the conception of physical necessity both inside and outside philosophy of science.[2]

I want to use the post-Russellian discussion of causation to argue that globalism is subtly but importantly mistaken. Here is my discussion plan: first I will say a little about why causal structure is not reducible to global laws, and then I will make a case for causal realism. After that, I will suggest that global laws are derivative from local laws that take the form of (what I will call) rules for mechanisms. Then, in the next section, I will argue that physics was never globalist, and finally I will talk about how rejecting globalism reopens the possibility of grounding causal structure in fundamental laws. Perhaps the most important upshot of the discussion is that it paves the way for an empiricist account of physical necessity, one that can be connected quite directly to experimental practice. I will conclude with some remarks about this.

THE CONTENT OF CAUSAL CLAIMS

Understanding the content of causal judgments has been a long, hard, and heavily contested road. For some time the philosophical discussion was dominated by attempts to provide analyses that systematize everyday intuitions about when *A* causes *B*. In recent years, something of a revolution occurred led by developments in cognitive and computer science, psychology, and statistics.[3] Instead of trying to systematize everyday intuitions about causes, attention turned to providing a formal framework for representing causal relations in science. We now have such a framework in the interventionist account of the content of causal claims that also sheds a good deal of light on everyday causal claims. The interventionist account came out of independent work by Glymour's group at Carnegie Mellon and Judea Pearl at the University of California–Los Angeles. Pearl's work culminated in his *Causality* (2000), though many people in the philosophical literature know interventionism from Woodward's *Making Things Happen* (2003a). I rely here on Pearl.

Pearl is a computer scientist and statistician, and he approached discussion of causal structure initially with the Bayesian presumptions that dominate his field. He writes:

In order to be combined with data, our knowledge must first be cast in some formal language, and what I have come to realize in the past ten years is that the language of probability is not suitable for the task; the bulk of human knowledge is organized around causal, not probabilistic relationships, and the grammar of probability calculus is insufficient for capturing those relationships. Specifically, the building blocks of our scientific and everyday knowledge are elementary facts such as "mud does not cause rain" and "symptoms do not cause disease" and those facts, strangely enough, cannot be expressed in the vocabulary of probability calculus. (2001, 19)

He set out to do for causal information what the probability calculus does for probabilities, supplementing the probability calculus with a formalism that was adequate to the expression of causal information. Pearl's goal was, in his words, "The enrichment of personal probabilities with causal vocabulary and causal calculus, so as to bring mathematical analysis closer to where knowledge resides" (2001, 19).

In Pearl's account, causes are to practical reasoning what probabilities are to epistemic reasoning. Whereas probabilities provide information about the correlations among a collection V of variables, causal information adds counterfactual information about *how changes in the value of one variable induce changes in the value of others.* Singular causal claims depend on generic causal information. Generic causal information is information about how one variable in a network induces changes in the values of other variables. Before we have a well-defined question about whether X_i is a cause of X_j, we have to specify a network. Once the network is specified, the question of whether X_i is a cause of X_j is the question of whether interventions on X_i induce changes in the value of X_j. An intervention on X_i is a change in its values that is "surgical" in the sense that it severs the connection between X_i and its parents in the network.[4] So if X_i is one of the variables in the network formed by V, knowing the causal effects of X_i is knowing what would happen if X_i were separated out of this web, severing connections with its own past causes and allowing it to vary. An "intervention" is just a formal name for the virtual act of separating a variable from its past causes.

Direct causation, represented by an arrow, is the most basic causal relation. A variable X_i is a direct cause of another variable X_j, relative to a variable set V, just in case there is an intervention on X_i that will change the value of X_j (or the probability distribution over the values of X_j) when all variables in V except X_i and X_j are held fixed (Pearl, 2001, 55). Causal relations are relative to networks. As variables are added to the network, new arrows appear, and others disappear. Causal relations of many kinds including total, direct, and indirect causes, necessary and sufficient causes, and actual and generic causes have been defined and successfully analyzed within the formal framework of structural causal models (SCM).

Scientists and philosophers of science have seized on Pearl's formalism for capturing the causal content of science. Psychologists and social scientists and econometricians have put it to work. The SCM framework is not an analysis of the folk concept of cause.[5] What it does rather is systematize the patterns of counterfactual judgments conveyed by everyday causal statements and needed to play the role identified by Cartwright in practical reasoning. What made Russell's eliminativism an insupportable position was not that science has to preserve folk intuitions about the world, but that causal reasoning and practical reasoning go hand in hand. Causal judgments supply the counterfactuals needed to identify strategic routes to action. When one is choosing between alternative courses of action, one needs to assess counterfactuals of the form "what *would* happen if I *A*'ed (rather than *B*'ed or *C*'ed)?" The builder choosing between wood and steel needs to know what would happen if he chose wood *and* what would happen if he chose steel, even though he will only do one or the other. The traveler choosing between path A and path B needs to know what would happen if he chose A *and* what would happen if he chose B, even though he cannot travel both. SCM shows us how to express claims of this form in exact logical terms and shows us why those claims are not captured by probabilities. As Pearl says,

> Probability theory deals with beliefs about an uncertain, yet static world, while causality deals with changes that occur in the world itself (or in one's theory of such changes). More specifically, causality deals with how probability functions change in response to influences (e.g., new conditions or interventions) that originate from outside the probability space, while probability theory, even when given a fully specified joint density function on all (temporally-indexed) variables in the space, cannot tell us how that function would change under such external influences. (2001, 36)

The interventionist analysis holds an important lesson for the proponents of the reductionist project. There is a very simple logical point that can be extracted from it that explains why the counterfactuals that we need to play the role of causal beliefs in deliberative reasoning will not in general be extractable from global dynamical laws. Let S be a system governed by a dynamical law L that gives the state of S at any time as a function of its state at earlier times. Let us suppose that it is a consequence of L that at any given time $A = kB$ (where k is a constant), and that there is some variable C downstream of A and B such that $C = f(A, B)$ (or more explicitly $C_t = f(A_{t-n}, B_{t-n})$, but I will suppress the temporal parameters).[6] There are three separate causal hypotheses not discriminated by the information that is given:

1. *A* causes *C*.
2. *B* causes *C*.
3. Neither *A* nor *B* cause *C*. *A*, *B*, and *C* are joint effects of a cause *D* in their common past.

Hypotheses 1, 2, and 3 are distinguished by the pattern of counterfactual dependence they postulate among *A*, *B*, and *C*. Hypothesis 1 entails that if *B* is held fixed and *A* is allowed to vary, *C* would vary as well. Hypotheses 2 and 3 entail that it would not. Hypothesis 2 entails that if *A* is held fixed and *B* is allowed to vary, *C* would vary as well. Hypotheses 1 and 3 entail that it would not. Hypothesis 3 entails that if *A* and *B* were held fixed and *D* were allowed to vary, *C* would vary as well. Hypotheses 1 and 2 entail that it would not.

The reason that the dynamical laws do not discriminate among these hypotheses is that the antecedents are counter*legals:* there are no models of the global laws in which the values of *A* and *B* vary independently.[7] In all of those models at all times, $A = kB$. Global laws underdetermine patterns of counterfactual dependence at the local level whenever there are interventions whose antecedents are not nomologically possible.[8] The information contained in a causal model is, in general, *strictly logically stronger* than the information contained in the global laws. Any law-like constraint on coevolution of local parameters is going to be preserved by evolution, and only the result of hypothetical interventions whose antecedents are counterlegals is going to separate the causal hypotheses.

THE CASE FOR CAUSAL REALISM

Let us pause to take stock. Russell observed that causal relations do not appear in a fundamental theory. He suggested that the notion of cause is a folk notion that has been superseded by global laws of temporal evolution and has no place in exact science. Cartwright observed that causal information plays an indispensable role in practical reasoning—that is, that the functional essence of causal beliefs is to supply the information about the results of hypothetical interventions needed for practical reasoning. This was codified in the interventionist analysis, which provided a precise formal framework for representing and investigating causal relationships,[9] and made it easy to see why causal information outruns the information generally contained in global dynamical laws.

To react to this situation, it will be useful to have a better intuitive feel for the relationship between causal facts and global laws. The real beauty of the SCM is that it gives insight into what grounds modal claims in science. Causal models are generalizations of the structural equations used in engineering, biology, economics, and social science. In a causal model, a complex system is represented as a modular collection of stable and autonomous components called "mechanisms." The behavior of each of these is represented as a function, and changes due to interventions are treated as local modifications of these functions. The dynamical law for the whole is recovered by assembling these in a configuration that imposes constraints on their relative variation. If we know how a complex system decomposes into mechanism, we know how interventions on the input to one mechanism propagate through the system. But since there are many ways of putting together mechanisms to get the same evolution at the global level, we cannot in general recover the causal information from the global dynamics.

Consider a complex mechanical system like a washing machine. We can model such a machine as a unit and write down an equation that allows us to calculate its state at one time from its state at any other: $[S_{final} = f(S_{initial})]$. If we were simply interested in description or prediction, and we knew the initial state of the engine, this would tell us everything there was to know. But if we want the kind of working knowledge that would let us troubleshoot, or intercede to modify the machine's behavior, f would not be enough. That kind of information is usually conveyed by a diagram that decomposes the machine into separable components, indicates how the components would behave if the parts were separated out and their input allowed to vary without constraint, and helps the viewer form an understanding of how the fixed connections among the parts within the context of the machine produce the overall pattern of behavior.[10] The global dynamics for the machine as a whole contain information about how its state varies over time, but it is the decomposition into mechanisms that tells us what would happen under interventions that do not arise under the normal evolution of the assembled machine.

In logical terms, the reason causal information goes missing when we just give the dynamics for the overall state is that the rules governing components are modally richer than the rule governing the whole. Embedded in a larger machine, the input to each component is constrained by the fixed connections among the parts of the engine so that each now moves within a restricted range and the patterns of counterfactual dependence that the interventionist sees as crucial to causal claims are lost. Variables that were allowed to vary freely in the original model are constrained by the values of variables in the embedding model, so information about what *would* happen *if* they were allowed to vary without constraint is lost. The dynamical law for the whole is not *incompatible* with the laws that govern the components. It is just that if we only look at the way that a system evolves as a whole, we lose information about the modal substructure. There is a real and absolute loss of modal information that occurs when one moves from a narrow-scope model of a subsystem to a wider scope model in which the input to the sub-model is constrained by the values of variables included in the wider model. Different ways of piecing together subsystems, with different implications for the results of local interventions, preserve the global dynamics.

REACTION: THE PEARL INVERSION

One might react to this situation by denying that there are modal facts over and above those that can be derived from global laws. In this view, one says that there is no fact of the matter about what would happen if some parameter were magically separated out and allowed to vary freely. There are two things to say about this. The first recapitulates Cartwright's response to Russellian eliminativism. Causal information is indispensable in practical reasoning. If we were simply interested in prediction, laws for the cosmos combined with information about initial conditions would tell us everything we need to know. But we face choices about how to act in the world, and we are interested in knowing how various ways of acting *would* play out. This is a way of saying that our actions are nodes in causal networks. Whenever I ask "what would happen if I *A*'ed rather the *B*'ed," I am asking for specifically causal information about the effects of *A*'ing. We care about causal information because we are not mere observers of nature but agents, and our actions have for us the status of interventions.[11] For purposes of predicting whether an engine will break down, it does not matter whether dirty oil causes or is merely a sign of impending engine breakdown.[12] It does not matter, that is to say, whether dirty oil is the symptom or the disease. The causal information matters for the mechanic who needs to know whether cleaning the oil will solve the problem.[13] We do not encounter the world as observers. We encounter it as agents. To act on the world, we need to know how it would respond to potential interventions.

The second thing to say questions the motive for rejecting causal information in favor of global laws. The components of the world are open systems. We encounter them severally, in multiple settings. Ideally we can isolate them in the laboratory and study their behavior individually and in interaction with other systems. Do we really think that the counterfactual implications of global laws purporting to describe completely specified alternatives to actuality are on a more secure epistemic or conceptual footing than the counterfactual implications of models of open subsystems of the actual world that we can isolate in the laboratory and study under conditions that approximate intervention?[14] Philosophers tend to be uninterested in partial views of bits of the world. They make a lunge for the most encompassing view. Most day-to-day science, however, is not concerned with the world as a unit, but is focused on local subsystems. The experimental scientist does his best to carve off a manageable bit of the universe. In the best case, his study is more or less tightly focused on a smaller unit, which can be isolated in the laboratory and whose responses to controlled interventions can be observed. That is not possible with larger systems, but we piece together an understanding of larger systems from an understanding of the rules governing components in constrained configuration.[15]

Models of open subsystems do have modal content: they identify counterfactual supporting regularities, so they involve induction from the observed results of actual interventions to merely potential ones. But the modal content is empirically grounded in testable regularities. When developing a model of an open subsystem, the scientist isolates the system as well as she can, identifies the variables whose causal effects she is interested in, finds some way of manipulating them while holding fixed the features of the internal configuration and environment she is imposing as constraints, and observes the effects. While there are practical difficulties in experimentally realizing situations to test for particular modal claims, there is nothing in principle untestable about modal claims pertaining to open subsystems of the universe.[16]

Things are different at the global level. Modal claims at that level purport to describe completely specified alternatives to actuality. They involve inductions from one case (the actual world) to a space of merely possible worlds. Possible worlds are entirely extrinsic to the actual world and are not even potentially observed. It is hard to say not only how we know about nonactual possible worlds but also why we care about them. The question of how an open system would behave if acted on in various ways, by contrast, has a transparent connection to actual things and an obvious practical interest. The question of how an open system would behave if acted on in various ways is what gives the idea of modality its practical significance. This idea gets generalized in causal models and applied to hypothetical interventions that go beyond what we can actually effect, but without losing its significance (in the same way that the idea of a view of the world from positions in space that we have no way of getting to makes good, if conjectural, sense). We get a problem if we reverse the order of explanation, reduce the local claims to global ones, then struggle to find an interpretation for these purportedly global modal facts. Unlike modal claims pertaining to open subsystems of the universe, which can be understood in terms of how those systems would respond to hypothetical interventions, the modal content of global models is *strictly* and *irremediably* counterfactual. Its semantic content is strictly extrinsic to the actual world.

In practice, we arrive at global laws like Newton's by extrapolation from the laws that govern its components. The modal implications of the global form of that law have no empirical or practical significance of their own. We are able to form beliefs about what would happen under hypothetical conditions because the world is composed of mechanisms that can be investigated independently and then recombined into larger systems whose behavior is a function of the rules governing components. The modal content trickles up from the experimentally based understanding of relatively simple components to larger configurations, rather than the other way around. It begins with modal generalizations that apply to the sorts of controlled subsystems that we can study in the laboratory. We form an understanding of larger units by piecing together what we know of the components: how they behave individually and in interaction when they are allowed to move freely and in constrained configurations. That is how knowledge obtained in the laboratory can lead to empirically well-founded beliefs about configurations that have not themselves been studied, and to empirically well-founded beliefs about configurations that have been studied to interventions that have not been observed. When we construct a new bridge or building, one that is not a copy of anything that has gone before, or when we synthesize a new pharmaceutical agent, we are not making wild inductive leaps of the kind we associate with theoretical breakthroughs. We are combining well-understood components in new ways. Every piece of new technology designed on paper that behaves as expected is possible because our modal knowledge is compositional in this way. When we understand how the components behave and the compositional principles, in a great many cases, we understand configurations.

In practice, the empirical content of laws attaches to predictions for open subsystems. These are derived not from global laws but from the rules pertaining to the mechanisms of which they are composed.

PHYSICS AND NATURE'S ULTIMATE MECHANISMS

One might argue that there is something quite misleading about these examples; we get this result only because in these examples we are tacitly restricting attention to global possibilities that leave the machine configuration intact. And the intuition that there is missing modal information is trading on the fact that we can imagine taking the machine apart and reconfiguring its parts. But if there are no law-like restrictions on the composition of mechanisms, then there is a global possibility for every local intervention and the difference disappears. So if we insist that in a well-behaved theory there is a global possibility that corresponds to every describable reconfiguration of components, then although there will not be an actual intervention that alters any of what we regard as the frozen accidents of our world (i.e., contingent features of initial conditions preserved by temporal evolution), there will be a model of the laws in which the antecedent of any intervention counterfactual holds.[17] Another way to put this is that if we are given the phase space for the universe as a whole and focus attention on the subspace that corresponds to the state of any subsystem of the world, treating everything else as exogenous, in a well-behaved theory (i.e., one in which there are no ultimate restrictions on configuration of components) it is arguable that there is always going to be a possible global state for every point on the boundary of the subspace. And if this is correct, then we can substitute global laws with a combinatorial property and capture the logical content of rules for mechanisms. So long as there is a global possibility for every local intervention, the laws pertaining to mechanisms will be recoverable from global laws.

In my view, this observation simply *reinforces* the point that our modal knowledge is rooted in our understanding of mechanisms. In a well-behaved theory, there is a global possibility for every local intervention because it is our ideas about what nature's basic mechanisms are that drive our ideas about what global configurations there are, rather than the other way around (*vide* Ismael 2013). What *makes* such a theory well behaved is that the global possibilities allow recombination of mechanisms.[18] But the observation does bring out a subtle ambiguity in the notion of law. Globalism is the thesis that the global laws of temporal evolution for our world are the most basic nomic generalizations. Those are laws that tell us how to calculate the state of our universe at one time from its state at other times. But there is another notion of law that is much more general, according to which global laws are laws that tell us not just how the state of our universe varies over time, but what kinds of universes are possible.[19] This is the kind of law that a fundamental theory in physics gives us. When such a theory is presented in physics, laws are given for simple components, and laws for complex systems are built up from those. Any way of piecing components together counts as a possible global configuration. We obtain the law of temporal evolution for our universe by specifying its initial configuration.[20]

Here is where I think we come back to the question of where Russell went wrong. When Russell looked for the most basic physical laws, he took the form of Newton's laws that apply to our world as a whole. What he *should* have done is taken the Newtonian laws governing the basic components of nature—nature's ultimate mechanisms—as basic. The fundamental law of temporal evolution for our world (the one for which Russell used the Newtonian law described in the quoted passage earlier) is the special form that these laws take for a system made up of the particular set of components of which our world is made arranged in a particular way. The composition of our world and its initial configuration encode contingent information needed to obtain detailed predictions. But the modal content of a theory is contained in the laws that govern components and the rules of composition. We cannot in general recover those from the global dynamics. So we have a trade-off in categorical and modal content. The rules for the components are weaker in categorical content than the law of temporal evolution for our world, but richer in modal content. The more categorical content we include, the less modal information we convey. To put it another way, the more information we have about what is actually the case, the less information about what would happen under non-actual conditions.

I remarked earlier that Pearl is not a metaphysician. He approached discussion of causation from the point of view of the statistician. In the preface to *Causality* (2000) he describes how his own thinking shifted away from the Bayesian presumptions that dominated his own field to the view that causal structure was fundamental:

[I used to think that] causality simply provides useful ways of abbreviating and organizing intricate patterns of probabilistic relationships. Today, my view is quite different. I now take causal relationships to be the fundamental building blocks both of physical reality and of human understanding of that reality, and I regard probabilistic relationships as but the surface phenomena of the causal machinery that underlies and propels our understanding of the world. (xiii–xiv)

I am suggesting a parallel shift away from the globalist presumptions that dominate the philosophy of science. I used to think that talk of mechanisms was a useful way of conveying partial information about global laws. Today, my view is quite different. I now take mechanisms to be the fundamental building blocks both of physical reality and of scientific understanding of that reality, and I regard global laws as but the emergent product of the mechanisms that underlie and propel our understanding of the world.

PHILOSOPHICAL IMPACT

This shift in thinking has several kinds of philosophical impact. In the first place, it reopens the possibility of grounding causal claims in fundamental law, suggesting a rather different research program for a physical fundamentalist. Instead of trying to derive causal facts from global laws, he sees the causal relations captured in directed acyclic graphs (DAGs) as emergent regularities rooted in composition of mechanisms.[21]

In the second place, issues about the metaphysics and epistemology of laws look rather different when rules for mechanisms are substituted for global laws. There is no question that science is steeped in modality. It studies not just what does happen but what could, and must, and would happen under hypothetical conditions. But the modal commitments of science create a dilemma for the empiricist. On one hand, belief in science seems the hallmark of empiricist commitment. On the other hand, believing that the world is governed by global laws, together with the inflated metaphysical commitments which that seems to carry, runs counter to the empiricist instinct.[22]

Directing our hermeneutic attention *away* from global laws toward the kinds of testable regularities pertaining to smaller than world-sized components of nature that we can isolate and study in the laboratory is a positive development for the empiricist. These have a well-behaved epistemology and make the most direct contact with our practical interests.[23] The empiricist can be discriminating about modal claims in science. She should not try to eliminate the modal content of a theory. The modal implications of theory play an indispensable role in guiding our interactions with open subsystems of the world and are grounded in empirically testable regularities. But she can be less tolerant of laws that pertain specifically to worlds as wholes.[24] It is only at the global level that modality becomes weird. To the extent to which they are not mere extrapolations of local modalities, global laws are no longer grounded in testable regularities. They become about other *worlds* rather than a hypothetical variation in our world, and they lose touch with the practical and empirical significance modal claims have for embedded agents. From an empiricist point of view, there is something altogether *upside down* about thinking that modal facts pertaining to the world as a whole are more epistemically or metaphysically secure than modal facts pertaining to open subsystems of the world. To take global laws as primitive and reject modal facts that cannot be reduced to them is to reject something that is immanent, empirically accessible, and metaphysically unmysterious in favor of something that is otherworldly, in principle inaccessible, and metaphysically exotic.[25]

I am strongly inclined to be a realist about modal claims grounded in testable rules for components. The fact that the universe is built up out of mechanisms we can separate from their environments in the laboratory and study in (approximate) isolation is what makes it possible to form modal beliefs. We should not try to reduce the modal implications of our theories, but we should try to ground them in rules for mechanisms, as inductions from testable regularities. I am strongly *dis*inclined, however, to be a realist about modal claims grounded in global laws. Where the globalist says, "Accept modal claims that can be derived from global laws, reject the overflow," I say, "Accept counterfactuals that can be derived from rules for mechanisms, reject the overflow." Or, more cautiously, leaving it open that there might be inductive practices that allow us to make modal inferences at the global level, I say that the burden of proof lies with the person invoking modal claims that cannot be grounded in rules for mechanisms to clarify their empirical basis.[26] What I am deeply suspicious of, however, is a tendency in foundational discussions to invoke global modal claims in an explanatory role. So, for example, in cosmology global laws are invoked to explain why certain global configurations do not arise. But to say that certain kinds of configurations cannot arise because the global laws rule them out strikes me as empty unless the global laws can be derived from some deeper principle.

Finally, this shift in how we think of laws, though in some ways a subtle shift when just thinking about how to express the modal content of science, can have a large impact at the hermeneutic level. Imaginative pictures guide first-order philosophical views, and the imaginative picture that comes with a globalist conception of laws is particularly toxic. The picture of natural necessity as deriving from global laws is very different from one that sees natural necessity as grounded in rules for mechanisms. Instead of ironclad global laws that seem to force history to unfold in one very particular way from its starting point, we have rules that describe the way that nature's simple components behave, something like the rules for chess pieces or the degrees of freedom and ranges of motion that define the behavior of the agitator and drum in the washing machine mentioned previously. These individual rules give rise to complex regularities when the components to which they pertain are placed in different configurations, which can be exploited by well-positioned agents who have control over parts of the machinery (or indeed who *are* parts of the machinery) to bring about more distal ends. The fact that our world is composed of simpler mechanisms that can be isolated and studied in the laboratory is what makes inductive practices and science possible. And that in its turn allows us to identify strategic routes to bringing about ends. It is what allows us to predict and control nature, and to gear our own actions toward desired ends. Laws and causes and all the inductive products of science are part of that. They are not (as the globalist picture encourages us to think) chains that bind us to act as we do. They are handmaids to choice.[27]

THIS CHAPTER HAS been advocating a return to a conception of modality grounded in scientific practice and an unwillingness to divorce science from experimental practice. I began with Russell's observation that causes have disappeared from the fundamental level of physical description, gave reasons for thinking that they were nevertheless indispensable for embedded agents, introduced the interventionist account as a formalization of the causal content of science, and showed why the modal content of causal claims generally outruns that of global laws. I then suggested that where Russell went wrong was in taking the global laws of temporal evolution modeled on the law of gravitation for our universe as the most basic nomic generalizations in science; I suggested instead that we take the rules that govern the behavior of nature's basic components as basic.[28]

In making this hermeneutic shift we do several things:

1. We reopen the possibility of recovering at least a large class of causal relationships (those captured in DAGs and formalized by SCM) as emergent regularities grounded in fundamental laws.
2. We clarify the epistemological basis of modal claims in science and pave the way for a moderate empiricist account of alethic modality. Modal judgments in science that can be rooted in rules for mechanisms rather than global laws are just inductions from testable regularities.
3. We free ourselves from an imaginative picture of laws that has played an insidious role outside philosophy of science.[29]

The fixation on global laws is part of a more general tendency among philosophers to focus on the products of science rather than the practice. The best antidote to that tendency is a focus on the experimental side of modeling. If we fixate on global laws, causal structure disappears and becomes difficult to recover. Modal generalizations seem metaphysically mysterious and detached from anything that can be observed. Because experimental practice is by its nature concerned with open subsystems of the world, this directs our gaze away from the global models and toward models of open subsystems. It allows us to connect modal generalizations to testable regularities, established in the laboratory by observing the results of interventions in a controlled setting. And by linking scientific modeling to intervention and manipulation, it gives us the tools for a more pragmatic approach to modal content.

NOTES

1. There are many excellent discussions of Russell's paper and Cartwright's response. See Field (2003) and the papers in Price and Corry (2007). A good deal of the post-Russellian discussion has focused on locating the source of the temporal asymmetry of causation. I will be focusing on a different issue, namely whether the modal content of causal claims can be grounded in fundamental law.

2. See, for example, the notion of laws at work in Helen Steward's discussion of freedom (2012). Her discussion is an unusually explicit but not atypical expression of the notion of physical law that many philosophers take from physics.

3. See Sloman (2005) for a user-friendly summary of these developments.

4. Formally, this amounts to replacing the equation governing X_i with a new equation $X_i = x_i$, substituting for this new value of X_i in all the equations in which X_i occurs, but leaving the other equations themselves unaltered. An intervention is defined as follows: "The simplest type of external intervention is one in which a single variable, say X_i, is forced to take on some fixed value x_i. Such an intervention, which we call 'atomic,' amounts to lifting X_i from the influence of the old functional mechanism $x_i = f_i(pa_i, u_i)$ and placing it under the influence of a new mechanism that sets the value x_i while keeping all other mechanisms unperturbed. Formally, this atomic intervention, which we denote by $do(X_i = x_i)$ or $do(x_i)$ for short, amounts to removing the equation $x_i = f_i(pa_i, u_i)$ from the model and substituting $X_i = x_i$ in the remaining equations" (Pearl, 2000, 70). So for Pearl, once you know what the causal mechanisms are, you can say which interactions constitute interventions. Woodward thinks that this limits the utility of interventions to discover causal mechanisms (among other things) and wants to characterize the notion of an intervention independently so that it can be used as a probe for causal structure. To some extent this in-house dispute reflects a difference in focus. From a metaphysical perspective, it is natural to take the underlying causal structure as basic. It is what explains the surface regularities and patterns of counterfactual dependence. But Woodward is interested in using interventions as a route in, so to speak. He wants to be able to identify interventions (perhaps provisionally) before we have a detailed understanding of the causal structure and use them to probe.

5. Although, see the Appendix to *Causality,* where Pearl has had a lot to say about why causal information is needed for deliberating, and connects it to the working knowledge that we associate with knowledge of how things work. The formalism is linked in this way with everyday notion of cause.

6. This sort of case arises routinely in medical situations in which a doctor needs to distinguish symptoms from cause.

7. One can see this difficulty in the acrobatics that possible worlds' semanticists face assessing counterfactuals that involve local departures from actuality.

8. We can make the same points by talking about phase spaces. When we develop a model of a constrained subsystem of the world, we restrict attention to a subspace of the global phase space. Knowing all the allowed trajectories through the global phase space will not give us the counterfactual

information we need to make causal judgments if there are phase points corresponding to free variation of local variables with no global trajectories through them.

9. Interventionism is an account of the content of causal claims, which we can take to mean an account of the inferential implications of causal beliefs, their role in epistemic and practical reasoning, and their relations to perception and action. For assessment of the SCM formalism and its impact on scientific investigation of cause, see Sloman (2005). For the philosophical development of interventionism and its relationship to alternatives, see Woodward (2003a, 2003b, 321–340).

10. One can look at a different level of resolution, add or subtract variables from the collection, or change background assumptions. Each of these constitutes a change in network and can alter causal relations among nodes. The causal relations at one level of resolution, relative to one collection of variables, and against a given set of background assumptions are different from those at another.

11. See Joyce (2007) and Ismael (2011) for discussion of decisions and their status.

12. The claim is not that it is generally irrelevant but that the specifically causal information does not add anything for predictive purposes to the probabilistic information, because the *specifically causal* information adds only information about unrealized possibilities.

13. Causal knowledge also matters for assigning responsibility for past events and learning from mistakes. It matters for understanding the significance of our choices and gauging their effects. It matters for deciding how to feel about the past and our role in it. Our emotional lives are built around "would have been and could have been."

14. For recent views on the elimination or reduction of causation to physical laws, see Norton (2007) and Maudlin (2007).

15. Sometimes nature creates a natural laboratory in the interaction between an open subsystem and its environment, but that is the exception rather than the rule.

16. There are general, effective ways of isolating a system causally from its environment, shielding it from the effects of exogenous variables. It is possible to operationalize interventions at the local level by inserting a random or pseudo-random process that fixes the value of exogenous variables. In universes in which there were no such processes, it would be impossible to discover causal structure, and science as we know it would be impossible.

17. There will not be interventions that change the energy or charge of a system, for example, just as there will not be interventions that alter accidental correlations among variables preserved by evolution.

18. The combinatorial principle is not a logical truth, and its status is contested. Hypotheses that violate it have irreducibly global constraints on configurations, and there are good reasons for insisting that global constraints should be emergent from rules governing the parts of which the world is composed. This is an issue that needs deeper examination.

19. Compare laws that tell us how a system of a given type evolves over time from any point in its phase space with laws that tell us how to construct the phase spaces for physically possible systems.

20. Where by "initial" we mean only "initial relative to a chosen interval." We do not mean some absolute initial moment in history. Finding the law of temporal evolution for our universe would be akin to specifying the Hamiltonian for our universe.

21. This is an open research program. Carrying it out would involve two components: (1) formalizing the full range of causal claims and supplementing the interventionist framework if it is not adequate to their expression, and (2) investigating whether this full range can be grounded in rules for mechanisms.

22. Van Fraassen famously regarded the rejection of modality as definitive of an empiricist stance toward science. Few have followed van Fraassen. For discussion of van Fraassen's view, see Ladyman (2000) and Monton and van Fraassen (2003).

23. The view that modal beliefs are beliefs about other possible worlds has infiltrated philosophy and distorted the content of modal belief in more complex ways that it would take more time to untangle.

24. The same remarks apply to other forms of modality defined over totalities, see, for example, Loewer and North (forthcoming).

25. One might say here that what I have given is an account of the source of modal belief but not modal fact. In my view, these are the same question.

26. These are the positive and negative sides of what Smolin calls "physics in a box" (2013).

27. See Ismael (2013).

28. Rules for configurations emerge from, and supervene on, those for components. Fix the rules for components and you fix the rules for configurations, but the converse is not true. Fix the rules for configurations, and you fix the rules for components only if we add a combinatorial principle that

guarantees that there is a global possibility for every describable configuration of components.

29. I have been reticent about attributing globalism to proponents of reductive projects because in almost every case the exposition is ambiguous between globalism and the closely related cousin discussed above, in this section. If the distinction is not made, or not made clearly enough, it is very easy to follow Russell's lead and think of the reductive project in globalist terms, handicapping the project beyond recovery. And the very small step from a globalist conception of law to a picture of the metaphysics of fundamental laws that has a wide currency outside philosophy of science has effects that ripple through philosophy.

REFERENCES

Cartwright, Nancy. 1979. "Causal Laws and Effective Strategies." *Nous* 13: 419–37.

Field, Hartry. 2003. "Causation in a Physical World." In *The Oxford Handbook of Metaphysics,* edited by Michael J. Loux and Dean Zimmerman, 435–60. Oxford: Oxford University Press.

Ismael, Jenann. 2011. "Decision and the Open Future." In *The Future of the Philosophy of Time,* edited by Adrian Bardon, 149–68. New York: Routledge.

Ismael, Jenann. 2013. "Causation, Free Will, and Naturalism." In *Scientific Metaphysics,* edited by Don Ross, James Ladyman, and Harold Kincaid, 208–35. Oxford: Oxford University Press.

Joyce, James. 2007. "Are Newcomb's Problems Decisions?" *Synthese* 156: 537–62.

Ladyman, James. 2000. " What's Really Wrong with Constructive Empiricism? Van Fraassen and the Metaphysics of Modality." *British Journal for the Philosophy of Science* 51: 837–56.

Loewer, Barry, and Jill North. Forthcoming. "Probing the Primordial Probability Distribution." In *Time's Arrows and the Probability Structure of the World,* edited by Barry Loewer, Brad Weslake, and Eric Winsberg. Cambridge, Mass.: Harvard University Press.

Maudlin, Tim. 2007. "Causation, Counterfactuals, and the Third Factor." In *The Metaphysics within Physics,* 104–42. Oxford: Oxford University Press.

Monton, Bradley, and Bas C. van Fraassen. 2003. "Constructive Empiricism and Modal Nominalism." *British Journal for the Philosophy of Science* 54: 405–22.

Norton, John. 2007. "Causation as Folk Science." In *Causation, Physics and the Constitution of Reality: Russell's Republic Revisited,* edited by Huw Price and Richard Corry, 11–44. Oxford: Oxford University Press.

Pearl, Judea. 2000. *Causality: Models, Reasoning, and Inference.* Cambridge: Cambridge University Press.

Pearl, Judea. 2001. "Bayesianism and Causality, or, Why I Am Only a Half-Bayesian." In *Foundations of Bayesianism,* edited by D. Corfield and J. Williamson, 19–36. Dordrecht, the Netherlands: Kluwer Academic.

Price, Huw, and Richard Corry, eds. 2007. *Causation, Physics and the Constitution of Reality: Russell's Republic Revisited.* Oxford: Oxford University Press.

Russell, Bertrand. (1913) 1953. "On the Notion of Cause," In *Mysticism and Logic,* 171–96. London: Doubleday.

Sloman, Steven. 2005. *Causal Models: How We Think about the World and Its Alternatives.* New York: Oxford University Press.

Smolin, Lee. 2013. *Time Reborn: From the Crisis in Physics to the Future of the Universe.* New York: Houghton Mifflin Harcourt.

Steward, Helen. 2012. *A Metaphysics for Freedom.* Oxford: Oxford University Press.

Woodward, James. 2003a. *Making Things Happen: A Theory of Causal Explanation.* Oxford: Oxford University Press.

Woodward, James. 2003b. "Critical Notice: Causality by Judea Pearl." *Economics and Philosophy* 19: 321–40.

7

EXPERIMENTAL FLUKES AND STATISTICAL MODELING IN THE HIGGS DISCOVERY

DEBORAH MAYO

By and large, Statistics is a prosperous and happy country, but it is not a completely peaceful one. Two contending philosophical parties, the Bayesians and the frequentists, have been vying for supremacy over the past two-and-a-half centuries . . . Unlike most philosophical arguments, this one has important practical consequences. The two philosophies represent competing visions of how science progresses.

—EFRON 2013

I$_F$ YOU SET OUT to explore the experimental side of discovering and testing models in science, your first finding is that experiments must have models of their own. These experimental models are often statistical. There are statistical data models to cope with the variability and incompleteness of data, and further statistical submodels to confront the approximations and idealizations of data–theory mediation. The statistical specifications and statistical outputs are scarcely immune to the philosophical debates of which Efron speaks—especially in cases of costly, Big Science projects.

Consider one of the biggest science discoveries in the past few years: the announcement on July 4, 2012, of evidence for the discovery of a Higgs-like particle based on a "5 sigma observed effect."[1] In October 2013, the Physics Nobel Prize was awarded jointly to François Englert and Peter W. Higgs for the "theoretical discovery of a mechanism" behind the particle discovered by the collaboration of thousands of scientists on the ATLAS (A Toroidal Large Hadron Collider ApparatuS) and CMS (Compact Muon Solenoid) teams at the Conseil Européen pour la Recherche Nucléaire (CERN) Large

Hadron Collider (LHC) in Switzerland. Because the 5-sigma standard refers to a benchmark from frequentist significance testing, the discovery was immediately imbued with controversies that, at bottom, concerned statistical philosophy.

Normally such academic disagreements would not be newsworthy, but the mounting evidence of nonreplication of results in the social and biological sciences as of late has brought new introspection into statistics, and a favorite scapegoat is the statistical significance test. Just a few days after the big hoopla of the Higgs announcement, murmurings could be heard among some Bayesian statisticians as well as in the popular press. Why a 5-sigma standard? Do significance tests in high-energy particle (HEP) physics escape the misuses of P values found in the social sciences and other sciences? Of course the practitioners' concern was not all about philosophy: they were concerned that their people were being left out of an exciting, lucrative, long-term project. But unpacking these issues of statistical-experimental methodology is philosophical, and that is the purpose of this discussion.

I am not going to rehash the Bayesian–frequentist controversy, which I have discussed in detail over the years (Mayo 1996, 2016; Mayo and Cox 2006; Mayo and Spanos 2006; Mayo and Spanos 2011). Nor will I undertake an explanation of all the intricacies of the statistical modeling and data analysis in the Higgs case. Others are much more capable of that, although I have followed the broad outlines of the specialists (e.g., Cousins 2017; Staley 2017). What I am keen to do, at least to begin with, is to get to the bottom of a thorny issue that has remained unsolved while widely accepted as problematic for interpreting statistical tests. The payoff in illuminating experimental discovery is not merely philosophical—it may well impinge on the nature of the future of Big Science inquiry based on statistical models.

The American Statistical Association (ASA) has issued a statement about P values (Wasserstein and Lazar 2016) to remind practitioners that significance levels are not Bayesian posterior probabilities; however, it also slides into questioning an interpretation of statistical tests that is often used among practitioners—including the Higgs researchers! Are HEP physicists, who happen to be highly self-conscious about their statistics, running afoul of age-old statistical fallacies? I say no, but the problem is a bit delicate, and my solution is likely to be provocative. I certainly do not favor overly simple uses of significance tests, which have been long lampooned. Rather, I recommend tests be reinterpreted. But some of the criticisms and the corresponding "reforms" reflect misunderstandings, and this is one of the knottiest of them all. My second goal, emerging from the first, is to show the efficiency of simple statistical models and tests for discovering completely unexpected ways to extend current theory, as in the case of the currently reigning standard model (SM) in HEP physics—while blocking premature enthusiasm for beyond the standard model (BSM) physics.

PERFORMANCE AND PROBABILISM

There are two main philosophies about the role of probability in statistical inference: *performance* (in the long run) and *probabilism*. Distinguishing them is the most direct way to get at the familiar distinction between frequentist and Bayesian philosophies.

The *performance philosophy* views the main function of the statistical method as controlling the relative frequency of erroneous inferences in the long run. For example, a frequentist statistical test (including significance tests), in its naked form, can be seen as a rule: whenever your outcome x differs from what is expected under a hypothesis H_0 by more than some value, say, x^*, you reject H_0 and infer alternative hypothesis H_1. Any inference can be in error, and probability enters to quantify how often erroneous interpretations of data occur using the method. These are the method's error probabilities, and I call the statistical methods that use probability this way *error statistical*. (This is better than frequentist, which really concerns the interpretation of probability.) The rationale for a statistical method or rule, according to its performance-oriented defenders, is that it can ensure that regardless of which hypothesis is true, there is a low probability of both erroneously rejecting H_0 as well as erroneously failing to reject H_0.

The second philosophy, probabilism, views probability as a way to assign degrees of belief, support, or plausibility to hypotheses. Some keep to a comparative report; for example, you regard data x as better evidence for hypothesis H_1 than for H_0 if x is more probable under H_1 than under H_0: $\Pr(x; H_1) > \Pr(x; H_0)$. That is, the *likelihood ratio* (LR) of H_1 over H_0 exceeds 1. Statistical hypotheses assign probabilities to experimental outcomes, and $\Pr(x; H_1)$ should be read: "the probability of data x computed under H_1." Bayesians and frequentists both use likelihoods, so I am employing a semicolon (;) here rather than conditional probability.

Another way to formally capture probabilism begins with the formula for conditional probability:

$$\Pr(H\,|\,x) = \frac{\Pr(H \text{ and } x)}{\Pr(x)}$$

$\Pr(H|x)$ is to be read "the probability of H given x." Note the use of the conditional bar (|) here as opposed to the semicolon (;) used earlier. From the fact that $P(H \text{ and } x) = P(x|H)P(H)$ and $P(x) = P(x|H)P(H) + P(x|\sim H)P(\sim H)$, we get Bayes's theorem. $\Pr(H|x) =$

$$\frac{\Pr(x\,|\,H)P(H)}{\Pr(x\,|\,H)P(H) + \Pr(x\,|\sim H)P(\sim H)}$$

$\sim H$, *the denial of H,* consists of alternative hypotheses H_i. $\Pr(H|x)$ is the posterior probability of H, while the $\Pr(H_i)$ the prior. In continuous cases, the summation is replaced by an integral.

Using Bayes's theorem obviously does not make you a Bayesian, and many who consider themselves Bayesians, especially nowadays, do not view inference in terms of Bayesian updating. Bayesian tests generally compare the posteriors of two hypotheses, reporting the odds ratio:

$$\frac{\Pr(H_0\,|\,x)}{\Pr(H_1\,|\,x)} = \frac{\Pr(x\,|\,H_0)\Pr(H_0)}{\Pr(x\,|\,H_1)\Pr(H_1)}$$

They often merely report the Bayes Factor against H_0: $\Pr(x; H_0)/\Pr(x; H_1)$.

The performance versus probabilism distinction corresponds roughly to the frequentist–Bayesian distinction, even though there are numerous different tribes practicing under the two umbrellas. Notably, for our purposes, some of us practicing under the frequentist (performance) umbrella regard good long-run performance as at most a necessary condition for a statistical method. Its necessity may be called the *weak repeated sampling principle.* "We should not follow procedures which for some possible parameter values would give, in hypothetical repetitions, misleading conclusions most of the time" (Cox and Hinkley 1974, 45–46). Restricting concern only to performance in the long run is often dubbed the *behavioristic approach* and is traditionally attributed to Neyman (of Neyman and Pearson statistics). In my view, there is also a need to show the relevance of hypothetical long-run repetitions (there may not be any actual repetitions) to the particular statistical inference.

Good performance alone fails to get at *why* methods work when they do—namely, to assess and control the stringency of the test at hand. This is the key to answering a burning question that has caused major headaches in frequentist foundations: why should low error rates matter to the appraisal of a particular inference? I do not mean to disparage the long-run performance goal. The Higgs experiments, for instance, relied on what are called triggering methods to decide which collision data to accept or reject for analysis. So 99.99 percent of the data must be thrown away! Here the concern is to control the proportions of hopefully useful collisions to analyze. But when it comes to inferring a particular statistical inference such as "there's evidence of a Higgs particle at a given energy level," it is not probabilism or performance we seek to quantify but *probativeness*. Error probabilities report on how *capable* a method is at avoiding incorrect interpretations of data, and with the right kind of testing reasoning this information may underwrite the correctness of a current inference.

Dennis Lindley and the Higgs Discovery

While the world was toasting the Higgs discovery, there were some grumblings back at the International Society of Bayesian Analysis (ISBA). A letter sent to the ISBA list by statistician Tony O'Hagan was leaked to me a few days after the July 4, 2012 Higgs announcement. Leading subjective Bayesian Dennis Lindley had prompted it. "Dear Bayesians," the letter began, "A question from Dennis Lindley prompts me to consult this list in search of answers. We've heard a lot about the Higgs boson."

> Why such an extreme evidence requirement? We know from a Bayesian perspective that this only makes sense if (a) the existence of the Higgs boson . . . has extremely small prior probability and/or (b) the consequences of erroneously announcing its discovery are dire in the extreme. (O'Hagan 2012)

Neither of these seemed to be the case in his opinion: "Is the particle physics community completely wedded to frequentist analysis? If so, has anyone tried to explain what bad science that is?" (O'Hagan 2012).

Bad science? It is not bad science at all. In fact, HEP physicists are sophisticated with their statistical methodology—they had seen too many bumps disappear. They want to ensure that before announcing the hypothesis H^*, "a new particle has been discovered," that H^* has been given a severe run for its money. Significance tests and cognate methods (confidence intervals) are the methods of choice here for good reason.

WHAT ARE STATISTICAL HYPOTHESES TESTS?

I am going to set out tests in a general manner that will let us talk about both simple Fisherian tests and Neyman-Pearson (NP) tests of the type we see in the Higgs inquiry. I will skip many aspects of statistical testing that are not germane to our discussion; those explications I give will occur as the story unfolds and on an as-needed basis. There are three basic components. There are (1) hypotheses and a set of possible outcomes or data, (2) a measure of accordance or discordance, $d(X)$, between possible answers (hypotheses) and data (observed or expected), and (3) an appraisal of a relevant probability distribution associated with $d(X)$.

(1) **Hypotheses.** A statistical hypothesis H, generally couched in terms of an unknown parameter θ, is a claim about some aspect of the process that might have generated the data, $x_0 = (x_1, \ldots, x_n)$, given in a model of that process—often highly idealized and approximate. The statistical model includes a general family of distributions for $X = (X_1, \ldots, X_n)$, such as a Normal distribution, along with its statistical parameters Θ, say, its mean and variance (μ, σ). We can represent the statistical model as $M_\theta(x) := \{ f(x; \theta), \theta \in \Theta \}$, where $f(x; \theta)$ is the probability distribution (or density), which I am also writing as $\Pr(x; H_i)$, and a sample space S.

In the case of the Higgs boson, there is a general model of the detector within which researchers define a "global signal strength" parameter μ such that H_0: $\mu = 0$ "corresponds to the background-only hypothesis and $\mu = 1$ corresponds to the SM [Standard Model] Higgs boson signal in addition to the background" (ATLAS Collaboration 2012c). The question at the first stage is whether there is enough of a discrepancy from 0 to infer something beyond the background is responsible; its precise value, along with other properties, is considered in the second stage. The statistical test may be framed as a one-sided test, where positive discrepancies from H_0 are sought:

$$H_0: \mu = 0 \text{ vs. } H_1: \mu > 0.$$

(2) **Distance function (test statistic) and its distribution.** A function of the data $d(X)$, the *test statistic* reflects how well or poorly the data x_0 accord with the hypothesis H_0, which serves as a reference point. Typically, the larger the value of $d(x_0)$, the farther the data are from what is expected under one or another hypothesis. The term *test statistic* is reserved for statistics whose distribution can be computed under the main or "test" hypothesis, often called the null hypothesis H_0 even though it need not assert

nullness.[2] It is the $d(x_0)$ that is described as "significantly different" from the null hypothesis H_0 at a level given by the P value.

The *P value* (or *significance level*) associated with $d(x_0)$ is the probability of a difference as large or larger than $d(x_0)$, under the assumption that H_0 is true:

$$\Pr(d(X) \geq d(x_0); H_0).$$

In other words, the P value is a distance measure but with this inversion: the farther the distance $d(x_0)$, the smaller the corresponding P value.

Some Points of Language and Notation

There are huge differences in notation and verbiage in the land of P values, and I do not want to get bogged down in them. Here is what we need to avoid confusion. First, I use the term "discrepancy" to refer to magnitudes of the parameter values like μ, reserving "differences" for observed departures, as, for example, values of $d(X)$. Capital X refers to the random variable, and its value is lowercase x. In the Higgs analysis, the test statistic $d(X)$ records how many *excess events* of a given type are "observed" in comparison to what would be expected from background alone, given in standard deviation or sigma units. We often want to use x to refer to something general about the observed data, so when we want to indicate the fixed or observed data, people use x_0.

We can write $\{d(X) > 5\}$ to abbreviate "the event that test T results in a difference $d(X)$ greater than 5 sigma." That is a statistical representation; in actuality such excess events produce a "signal-like" result in the form of bumps off a smooth curve, representing the background alone. There is some imprecision in these standard concepts. Although the P value is generally understood as the fixed value, we can also consider the P value as a random variable: it takes different values with a given probability. For example, in a given experimental model, the event $\{d(X) > .5\}$ is identical to $\{P(X) < .3\}$.[3] The set of outcomes leading to a .5 sigma difference equals the set of events leading to a P value of .3.

(3) Test Rule T and its associated error probabilities. A simple significance test is regarded as a way to assess the compatibility between the data and H_0, taking into account the ordinary expected variability "due to chance" alone. Pretty clearly if there is a high probability of getting an even larger difference than you observed due to the expected variability of the background alone, then the data do not indicate an incompatibility with H_0. To put it in notation,

If $\Pr(d(X) > d(x); H_0)$ is high, then

$d(x)$ does not indicate incompatibility with H_0.

For example, the probability of $\{d(X) > .5\}$ is approximately .3. That is, a $d(X)$ greater than .5 sigma occurs 30 percent of the time due to background fluctuations under H_0. We would not consider such a small difference as incompatible with what is expected due to chance variability alone, because chance alone fairly frequently would produce an even *greater* difference.

Furthermore, if we were to take a .5 sigma difference as indicating incompatibility with H_0 we would be wrong with probability .3. That is, .3 is an *error probability* associated with the test rule. It is given by the probability distribution of $d(X)$—its *sampling distribution*. Although the P value is computed under H_0, later we will want to compute it under discrepancies from 0.

The error probabilities typically refer to hypotheticals. Take the classic text by Cox and Hinkley:

> For given observations x we calculate . . . the *level of significance* p_{obs} by
>
> $$p_{obs} = \Pr(d \geq d_0; H_0).$$
>
> . . . Hence p_{obs} is the probability that we would mistakenly declare there to be evidence against H_0, were we to regard the data under analysis as just decisive against H_0. (Cox and Hinkley 1974, 66; replacing their t with d, t_{obs} with d_0)

Thus, p_{obs} would be the test's probability of committing a type I error—erroneously rejecting H_0. Because this is a continuous distribution, it does not matter if we use > or $\geq$ here.

NP tests will generally fix the P value as the cutoff for rejecting H_0, and then seek tests that also have a high probability of discerning discrepancies—high *power*.[4] The difference that corresponds to a P value of α is d_α. They recommend that the researcher choose how to balance the two. However, once the test result is at hand, they advise researchers to report the observed P value, p_{obs}—something many people overlook. Consider Lehmann, the key spokesperson for NP tests, writing in Lehmann and Romano (2005):

> It is good practice to determine not only whether the hypothesis is accepted or rejected at the given significance level, but also to determine the smallest significance level . . . at which the hypothesis would be rejected for the given

observation. This number, the so-called *p-value* gives an idea of how strongly the data contradict the hypothesis. It also enables others to reach a verdict based on the significance level of their choice. (63–64)

Some of these points are contentious, depending on which neo-Fisherian, neo-Neyman-Pearsonian, or neo-Bayesian tribe you adhere to, but I do not think my arguments depend on them.

Sampling Distribution of d(X)

In the Higgs experiment, the probability of the different $d(X)$ values, the *sampling distribution* of $d(X)$, is based on simulating what would occur under H_0, in particular, the relative frequencies of different signal-like re-sults or "bumps" under H_0: $\mu = 0$. These are converted to corresponding probabilities under a standard normal distribution. The test's error prob-abilities come from the relevant sampling distribution; that is why *sam-pling theory* is another term to describe error statistics. Arriving at their error probabilities had to be fortified with much background regarding the assumptions of the experiments and systematic errors introduced by extraneous factors. It has to be assumed the rule for inference and "the re-sults" include all of this cross-checking before the P value can be validly computed.

The probability of observing results as or more extreme as 5 sigmas, under H_0, is approximately 1 in 3,500,000! It is essentially off the charts. Figure 7.1 shows a blow-up of the area under the normal curve to the right of 5.

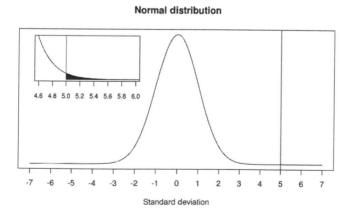

Normal distribution

Standard deviation

Figure 7.1. Normal distribution with details of the area under the normal curve to the right of 5.

Typically a P value of only .025 or .001 is required to infer an indication of a genuine signal or discrepancy, corresponding to 2 or 3 sigma, respectively. Why such an extreme evidence requirement? Lindley asked. Given how often bumps disappear, the rule for interpretation, which physicists never intended to be strict, is something like if $d(X) \geq 5$ sigma, infer discovery; if $d(X) \geq 2$ sigma, get more data. Clearly, we want methods that avoid mistaking spurious bumps as real—to avoid type 1 errors—while ensuring high capability to detect discrepancies that exist—high power. If you make the standard for declaring evidence of discrepancy too demanding (i.e., set the type 1 error probability too low), you will forfeit the power to discern genuine discrepancies. The particular balance chosen depends on the context and should not be hard and fast.

What "the Results" Really Are: Fisher's Warning

From the start, Fisher warned that in order to use P values to legitimately indicate incompatibility, we need more than a single isolated low P value. We need to demonstrate an *experimental phenomenon*.

> We need, not an isolated record, but a reliable method of procedure. In relation to the test of significance, we may say that a phenomenon is experimentally demonstrable when we know how to conduct an experiment which will rarely fail to give us a statistically significant result. (Fisher 1947, 14)

The interesting thing about this statement is that it cuts against the idea that nonsignificant results may be merely ignored or thrown away (although sometimes Fisher made it sound that way). You would have to be keeping track of a set of tests to see if you had a sufficiently reliable test procedure or to check that you had the ability to conduct an experiment that rarely fails to give a significant result. Even without saying exactly how far your know-how must go before declaring "a phenomenon is experimentally demonstrable," some reliable test procedure T is needed.

Fisher also warned of that other bugbear that is so worrisome these days: reporting your results selectively. (He called it "the political principle" [1955, 75].) That means it is not enough to look at $d(X)$; we have to know whether a researcher combed through many differences and reported only those showing an impressive difference—that is, engaged in data dredging, cherry-picking, and the like. Lakatos might say, we need to know the heuristic by which claims are arrived at.

Statistical tests in the social sciences are often castigated not only for cherry-picking to get a low P value (also called P hacking), but also for taking a statistically significant effect as automatic evidence for a research claim. The actual probability of erroneously inferring an effect is high. Something similar can occur in HEP physics under the "look elsewhere effect" (LEE), which I will come back to. We may capture this in a principle:

> **Severe Testing Principle (weak):** An observed difference $d(x)$ provides poor grounds to reject H_0 if, thanks to cherry-picking, P hacking, or other biasing selection effects, a statistically significant difference from H_0 would probably be found, even if H_0 were an adequate description of the process generating the data.

To put it in Popperian terms, cherry-picking ensures that H_0 probably would not have "survived," even if true or approximately true. You report it would be very difficult (improbable) to have obtained such a small P value assuming H_0, when in fact it would be very easy to have done so! You report what is called the *nominal P* value, in contrast to the *actual* one.

So, really, the required "results" or "evidence" include demonstrating the know-how to generate differences that rarely fail to be significant, and showing the test is not guilty of selection biases or violations of statistical model assumptions. To emphasize this, we may understand the results of Test T as incorporating the entire display of know-how and soundness. That is what I will mean by $\Pr(\text{Test } T \text{ displays } d(X) \geq d(x_0); H_0) = P$.

> **Severity Principle (strong):** If $\Pr(\text{Test } T \text{ displays } d(X) \geq d(x_0); H_0) = P$, for a small value of P, then there is evidence of a genuine experimental effect or incompatibility with H_0.

Let d_α be the difference that reaches the P value of a. Then

$$\Pr(\text{Test T displays } d(X) \geq d_\alpha; H_0) = \alpha$$

If I am able to reliably display a very statistically significant result without biasing selection effects, I have evidence of a genuine experimental effect.

There are essentially two stages of analysis. The first stage is to test for a genuine Higgs-like particle; the second is to determine its properties (production mechanism, decays mechanisms, angular distributions, etc.). Even though the SM Higgs sets the signal parameter to 1, the test is going to be used to learn about the value of any discrepancy from 0. Once the null at the first stage is rejected, the second stage essentially shifts to learning the particle's properties and using them to seek discrepancies with a new null hypothesis: the SM Higgs.

THE PROBABILITY OUR RESULTS ARE STATISTICAL FLUKES (OR DUE TO CHANCE VARIABILITY)

The July 2012 announcement about the new particle gave rise to a flood of buoyant, if simplified, reports heralding the good news. This gave ample grist for the mills of P value critics—whom statistician Larry Wasserman (2012) playfully calls the "P-Value Police"—such as Sir David Spiegelhalter (2012), a professor of the Public's Understanding of Risk at the University of Cambridge. Their job is to examine whether reports by journalists and scientists can be seen to be misinterpreting the sigma levels as posterior probability assignments to the various models and claims: thumbs-up or thumbs-down! Thumbs-up went to the ATLAS group report:

> A statistical combination of these channels and others puts the significance of the signal at 5 sigma, meaning that *only one experiment in three million would see an apparent signal this strong in a universe without a Higgs.* (ATLAS Collaboration, 2012a, italics added)

Now HEP physicists have a term for an apparent signal that is actually produced due to chance variability alone: *a statistical fluke.* Only one experiment in 3 million would produce so strong a fluke. By contrast, Spiegelhalter gave thumbs-down to the U.K. newspaper *The Independent*, which reported,

> There is less than a one in 3 million chance that their results are a statistical fluke.

If they had written "would be" instead of "is" it would get thumbs up. Spiegelhalter's ratings are generally echoed by other Bayesian statisticians. According to them, the thumbs-down reports are guilty of misinterpreting the P value as a posterior probability on H_0.

A careful look shows this is not so. H_0 does not say the observed results are due to background alone; H_0 does not say the result is a fluke. It is just H_0: $\mu = 0$. Although if H_0 were true it follows that various results would occur with specified probabilities. In particular, it entails (along with the rest of the background) that large bumps are improbable. It may in fact be seen as an ordinary error probability. Because it is not just a single result but a dynamic test display, we can write it:

(1) $\Pr(\text{Test } T \text{ displays } d(X) \geq 5; H_0) \leq .0000003$.

The portion within the parentheses is how HEP physicists understand "a 5 sigma fluke."

Note that (1) is not a conditional probability, which involves a prior probability assignment to the null. It is not

$\Pr(\text{Test } T \text{ displays } d(X) > 5 \text{ and } H_0) / \Pr(H_0)$

Only random variables or their values are conditioned upon. Calling (1) a "conditional probability" is not a big deal. I mention it because it may explain part of the confusion here. The relationship between the null and the test results is intimate: the assignment of probabilities to test outcomes or values of $d(X)$ "under the null" may be seen as a tautologous statement.

Critics may still object that (1) only entitles saying, "There is less than a one in 3 million chance of a fluke (as strong as in their results)." These same critics give a thumbs-up to the construal I especially like:

> The probability of the background alone fluctuating up by this amount or more is about one in three million. (CMS Experiment 2012)

But they would object to assigning the probability to the particular results being due to the background fluctuating up by this amount. Let us compare three "ups" and three "downs" to get a sense of the distinction.

Ups
U-1. The probability of the background alone fluctuating up by this amount or more is about 1 in 3 million.
U-2. Only one experiment in three million would see an apparent signal this strong in a universe described in H_0.
U-3. The probability that their signal would result by a chance fluctuation was less than one chance in 3.5 million.

Downs
D-1. The probability their results were due to the background fluctuating up by this amount or more is about 1 in 3 million.

D-2. One in 3 million is the probability the signal is a false positive—a fluke produced by random statistical fluctuation.

D-3. The probability that their signal was a result of a chance fluctuation was less than 1 chance in 3 million.

The difference is that the thumbs-down allude to "this" signal or "these" data being due to chance or being a fluke. You might think that the objection to "this" is that the P value refers to a difference as great or greater—a tail area. But if the probability $\{d(X) > d(x)\}$ is low under H_0, then $\Pr(d(X) = d(x); H_0)$ is even lower. (As we will shortly see, the "tail area" plays its main role with insignificant results.) No statistical account recommends going from improbability of a point result on a continuum under H to rejecting H. The Bayesian looks to the prior probability in H and its alternatives. The error statistician looks to the general procedure. The notation $\{d(X) > d(x)\}$ is used to signal a reference to a general procedure or test. (My adding "Test T displays" is to bring this out further.)

But then, as the critic rightly points out, we are not assigning probability to this particular data or signal. That is true, but that is the way frequentists always give probabilities to events, whether it has occurred or we are contemplating an observed excess of 5 that might occur. It is always treated as a generic type of event. We are never considering the probability "the background fluctuates up this much on Wednesday July 4, 2012," except as that is construed as a type of collision result at a type of detector, and so on.

Now for a Bayesian, once the data are known, they are fixed; what is random are an agent's beliefs, usually cashed out as their inclination to bet or otherwise place weights on H_0. So from the Bayesian perspective, once the data are known, $\{d(X) \geq 5\}$ is illicit, as it considers outcomes other than the ones observed. So if a Bayesian is reading D-1 through D-3, it appears the probability must be assigned to H_0. By contrast, for a frequentist, the error probabilities given by P values still hold post-data. They refer to the capacity of the test to display such impressive results as these.

The Real Problem with D-1 through D-3

The error probabilities in U-1 through U-3 are quite concrete: we know exactly what we are talking about and can simulate the distribution. In the Higgs experiment, the needed computations are based on simulating the relative frequencies of events where H_0: $\mu = 0$ (given a detector model). In terms of the corresponding P value,

(1) Pr(test T displays a P value $\leq$ .0000003; H_0) < .0000003

References to "their results" and "their signal" are understood the same way. So what is the objection to D-1, D-2, and D-3? It is the danger of moving from such claims to mistaken claims about their complements: If I say there is a .0000003 probability the results are due to chance, some people infer there is a .999999 (or whatever) probability their results are not due to chance—are not a false positive, are not a fluke. And those claims are wrong. If the Pr(A; H_0) = P, for some assertion A, the probability of the complement is Pr(not-A; H_0) = 1–P. In particular,

(1) Pr(test T would *not* display a P value $\leq$ .0000003; H_0) > .9999993.

There's no transposing! That is, the hypothesis after the semicolon does not switch places with the event to the left of the semicolon! But despite how the error statistician hears D-1 through D-3, I am prepared to grant that the corresponding U claims are safer. Yet I assure you my destination is not merely refining statistical language but something deeper, and I am about to get to that.

Detaching Inferences from the Evidence

Phrases such as "the probability that our results are a statistical fluctuation (or fluke) is very low" are common enough in HEP physics, although physicists tell me it is the science writers who reword their correct U claims as slippery D claims. Maybe so. But if you follow the physicist's claims through the process of experimenting and modeling, you find they are alluding to proper error probabilities. You may think they really mean an illicit posterior probability assignment to "real effect" or H_1 if you think that statistical inference takes the form of probabilism. In fact, if you are a Bayesian probabilist, and assume the statistical inference must have a posterior probability or a ratio of posterior probabilities or a Bayes factor, you will regard U-1 through U-3 as legitimate but irrelevant to inference, and D-1 through D-3 as relevant only by misinterpreting P values as giving a probability to the null hypothesis H_0.

If you are an error statistician (whether you favor a behavioral performance or a severe probing interpretation), even the correct claims U-1 through U-3 are not statistical inferences! They are the (statistical) justifications associated with implicit statistical inferences, and even though HEP practitioners are well aware of them, they should be made explicit. Such inferences can take many forms, such as those I place in brackets:

U-1. The probability of the background alone fluctuating up by this amount or more is about 1 in 3 million.

[Thus, our results are not due to background fluctuations.]

U-2. Only one experiment in 3 million would see an apparent signal this strong in a universe where H_0 is adequate.

[Thus, H_0 is not adequate.]

U-3. The probability that their signal would result by a chance fluctuation was less than 1 chance in 3.5 million.

[Thus, the signal was not due to chance.]

The formal statistics move from

(1) Pr(test T produces $d(X) \geq 5$; H_0) < .0000003.

to

(2) There is strong evidence for
(first) (2a) a genuine (non-fluke) discrepancy from H_0.
(later) (2b) H^*: a Higgs (or a Higgs-like) particle.

They move in stages from indications, to evidence, to discovery. Admittedly, moving from (1) to inferring (2) relies on an implicit assumption or principle of error statistical testing.

Testing Principle (from low P value): Data provide evidence for a genuine discrepancy from H_0 (just) to the extent that H_0 would (very probably) have survived were H_0 a reasonably adequate description of the process generating the data.

It is one instance of our general *severity principle,* and, as we have seen, it can be stated in many ways and has a number of corollaries. What *is* the probability that H_0 would have "survived" (and not been rejected) at the 5 sigma level? It is the probability of the complement of the event $\{d(X) \geq 5\}$, namely $\{d(X) < 5\}$ computed under H_0. Its probability is correspondingly 1–.0000003 (i.e., .9999997). So the overall argument, fleshed out in terms of the display of bumps, goes like this:

(1′) With probability .9999997, the bumps would be smaller, would behave like flukes (disappear with more data, would not be displayed at both CMS and ATLAS, etc.), in an experimental world adequately modeled by H_0.

They did not disappear; they grew (from 5 to 7 sigma). So (2a) infers there is evidence of H_1: non-fluke, and (2b) infers H^*: a Higgs (or a Higgs-like) particle.

The associated error probability qualifies the inference in (2). It is a report of the stringency or severity of the test that the claim has passed, as given in (1'): .9999997. Without making the underlying principle of testing explicit, some critics assume the argument is all about the reported P value. In fact, that is a mere stepping stone to an inductive inference that is detached.

A Performance-Oriented Interpretation

Someone who insisted on a strict NP performance construal might reason as follows. If you follow the rule of behavior and interpret 5 sigma bumps as a real effect (a discrepancy from 0), you would erroneously interpret data with probability less than .0000003—a very low *error probability*. Doubtless, HEP physicists are keen to avoid repeating such mistakes—such as apparently finding particles that move faster than light only to discover some problem with the electrical wiring (Reich 2012). But I claim the specific evidential warrant for the 5 sigma Higgs inferences are not low long-run errors, but being able to detach an inference based on a highly stringent test, or what is often called a *strong argument from coincidence*.[5]

Ian Hacking: There Is No Logic of Statistical Inference

The supposition that statistical inference must follow some form of probabilism is often the basis for criticizing frequentist tests for using an error probability to qualify the method. This was the basis for Ian Hacking's 1965 criticism of NP tests. In a fascinating article in 1980, Ian Hacking announced he was retracting his earlier criticism of NP tests. He now said he was wrong to claim it failed to provide an account of statistical inference. In fact, he announced, "There is no such thing as a logic of statistical inference" (Hacking 1980, 145), and furthermore, "I now believe that Neyman, Peirce, and Braithwaite were on the right lines to follow in the analysis of inductive arguments" (1980, 141). I agree with Hacking when he agreed with C. S. Peirce, who said,

> In the case of analytic [deductive] inference we know the probability of our conclusion (if the premises are true), but in the case of synthetic [inductive] inferences we only know the degree of trustworthiness of our proceeding. (2.693)

In getting new knowledge, in ampliative, or inductive reasoning, the conclusion should go beyond the premises; and probability enters to qualify the overall reliability of the method, or, as I prefer, the severity with which the inference has passed a test.

LEARNING HOW FLUKES BEHAVE: THE GAME OF BUMP-HUNTING

To grasp the trustworthiness of the proceeding or the severity of the overall testing, we must go beyond formal statistics. Let us get a sense of some of the threads of the background considerations that enter in the case of the Higgs. Dennis Overbye wrote an article in *The New York Times* on "Chasing the Higgs Boson," based on his interviews with spokespeople Fabiola Gianotti (ATLAS) and Guido Tonelli (CMS). It is altogether common, Tonelli explains, that the bumps they find are "random flukes"—spuriously significant results—"So 'we crosscheck everything' and 'try to kill' any anomaly that might be merely random." It is clear from Overbye's "bump hunting" story how the phrase "the chances the bumps are flukes" is intended.

> One bump on physicists' charts . . . was disappearing. But another was blooming like the shy girl at a dance. In retrospect, nobody could remember exactly when she had come in. But she was the one who would marry the prince . . . It continued to grow over the fall until it had reached the 3-sigma level—*the chances of being a fluke* [spurious significance] *were less than 1 in 740, enough for physicists to admit it to the realm of "evidence" of something, but not yet a discovery.* (Overbye 2013, italics added)

One time they were convinced they had found evidence of extra dimensions of space-time "and then the signal faded like an old tired balloon" (Overbye 2013). They had written it up in a paper, but fortunately withheld its publication. "We've made many discoveries," Dr. Tonelli said, "most of them false" (Overbye 2013).

What is the difference between HEP physics and fields where "most research findings are false" (Ioannidis 2005) or so we keep hearing? For one thing, HEP physicists do not publish on the basis of a single, isolated "nominal" *P* value. For another thing, they take into account what is known as the Look Elsewhere Effect (LEE). To explain, the null hypothesis is formulated to correspond to regions where an excess or bump is found. Not knowing the mass region in advance means "the local p value did not include the fact that 'pure chance' has lots of opportunities . . . to provide an unlikely occurrence" (Cousins 2017, 424). So here a local (like the nominal) P value is assessed at a particular, data-determined mass. But the probability of so impressive a difference *anywhere in a mass range—the global P value*—would be greater than the local one. Thus, "the original concept of '5σ' in HEP was therefore mainly motivated as a (fairly crude) way to account for a multiple trials factor . . . known as the 'Look Elsewhere Effect'" (Cousins 2017, 425). HEP physicists often report both local and global P values.

Background information enters not via prior probabilities of the particles' existence but as how one might be led astray. "If they were flukes, more data would make them fade into the statistical background. If not, the bumps would grow in slow motion into a bona fide discovery" (Overbye 2013). So physicists give the bump a hard time; they stress test, look at multiple decay channels, and they hide the details of where they found it from the other team. They deliberately remain blind to the mass area until the last minute, when all is revealed. When two independent experiments find the same particle signal at the same mass, it helps to overcome the worry of multiple testing (LEE), enabling a reasonably strong argument from coincidence.

Using Insignificant Results to Set Upper Bounds to the Discrepancy: Curb Your Enthusiasm

Once the null hypothesis is rejected, the job shifts to *testing if various parameters agree with the SM predictions*. Now the corresponding null hypothesis, call it H_0^2, is the SM Higgs boson,

$$H_0^2: \text{SM Higgs boson: } \mu = 1$$

and discrepancies from it are probed, often estimated with confidence intervals.

Here, I follow Robert Cousins (2017). This takes us to the most important role served by statistical significance tests—when the results are insignificant or the P values are not small. These are negative results. They afford a standard for blocking inferences otherwise made too readily. In this episode, they arose to

(a) deny sufficient evidence of a new particle,
(b) rule out values of various parameters, for instance, spin values that would preclude its being "Higgs-like".

Although the popular press highlighted the great success for the SM, the HEP physicists at both stages were vigorously, desperately seeking to uncover BSM (beyond the standard model) physics. Statistically significant results (very low P values) are the basis to infer genuine effects; whereas negative results (moderate P values) are used to deny evidence of genuine effects.

Once again, the background knowledge of fluke behavior was central to curbing the physicists' enthusiasm about bumps that apparently indicated discrepancies with

$$H_0^2 : \mu = 1$$

Even though the July 2012 data gave evidence of the existence of a Higgs-like particle—where calling it "Higgs-like" still kept the door open for an anomaly with the "plain vanilla" particle of the SM—they also showed hints of such an anomaly.

Matt Strassler, who like many is longing to find evidence for BSM physics, was forced to concede: "The excess [in favor of BSM properties] . . . *has become a bit smaller each time . . . That's an unfortunate sign, if one is hoping the excess isn't just a statistical fluke*" (2013a, italics added). Notice, the "excess" refers to an observed difference $d(x)$. So assigning probabilities to it is perfectly proper. Or they would see the bump at ATLAS . . . and not CMS. "Taking all of the LHC's data, and not cherry picking . . . there's nothing here that you can call 'evidence,'" for the much sought BSM (Strassler 2013b).

Considering the frequent flukes, and the hot competition between the ATLAS and CMS to be first, a tool to show when to curb their enthusiasm is exactly what is wanted. This "negative" role of significance tests is crucial for denying BSM anomalies are real, and setting the upper bounds for these discrepancies with the SM Higgs. Because each test has its own test statistic, I will use $g(x)$ rather than $d(x)$. A corollary of the same test principle (severity principle) now takes the following form.

> **Testing Principle (for nonsignificance):** Data provide evidence to rule out a discrepancy δ^* to the extent that a larger $g(x_0)$ would very probably have resulted if δ were as great as δ^*.
>
> In other words, *if* $Pr(g(X) > g(x_0))$ *is high, under the assumption of a genuine discrepancy δ^* from H_0^2, then $g(x_0)$ indicates $\delta < \delta^*$.*

I do not want to get too bogged down into the statistical notation, especially as the particular value of δ^* is not so important at this stage. What actually happens with negative results is that the discrepancies that are indicated get smaller and smaller, as do the bumps, and just vanish. This is one of the cases, in other words, where negative results lead to inferring no difference from the SM Higgs in the respect probed.[6]

Negative results in HEP physics are scarcely the stuff of file drawers, a serious worry that leads to publication bias in many fields. Cousins (2017, 412) tells of the wealth of articles that begin "search for . . ." These studies are regarded as important and informative—if only for ruling out avenues for theory development. Once one attempt at BSM is knocked out, everyone looks to another. (I will return to this in the 2016 update.) There is another lesson for domains confronted with biases against publishing negative results.

HOW DOES THE TRADITIONAL BAYESIAN MODEL OF EXPERIMENT FARE?

Returning to Tony O'Hagan's thoughts on the particle physics community, he published a digest of responses a few days later (O'Hagan 2012).[7] When it was clear his letter had not met with altogether enthusiastic responses, he backed off, claiming he had only meant to be provocative with the earlier letter. Still, he declared, the Higgs researchers would have been better off avoiding the "ad hoc" 5 sigma by doing a proper (subjective) Bayesian analysis. "They surely would be willing to announce SM Higgs discovery if they were 99.99 percent certain of the existence of the SM Higgs." Would it not be better to affirm

$$\Pr(\text{SM Higgs}|\text{data}) = .9999?$$

Actually, no. Physicists believed in the SM Higgs before building the big billion-dollar collider. Given the perfect predictive success of the SM and its simplicity, such beliefs would meet the familiar standards for plausibility. But that is very different than having evidence for a discovery, or information about the characteristics of the particle. Many apparently did not expect it to have so small a mass, 125 GeV. In fact, given the untoward consequences of this low mass, those researchers may well have gone back and changed their prior probabilities to arrive at something more "natural." But their strong arguments from coincidence via significance tests prevented the effect from going away.

O'Hagan and Lindley admit that a subjective Bayesian model for the Higgs would require prior probabilities to a plethora of high dimensional "nuisance" parameters of the background and the signal; it would demand multivariate priors, correlations between parameters, joint priors, and the ever worrisome Bayesian catchall: Pr(data|not H^*). Lindley's idea of subjectively eliciting beliefs from HEP physicists seems wholly unrealistic here. Many Bayesians, in practice, not only find subjective elicitation unreliable, they find it detracts from the more serious problem of model specification (Berger 2006). Merely explaining some of the statistical concepts, such as partial correlations of nuisance parameters, takes considerable time. No wonder the most prevalent type of Bayesian approach these days is based on one or another *default* or reference prior: they are formal conventions for replacing unknown parameters with priors and integrating them out.

An early leader was Harold Jeffreys. Jim Berger (2006) recommends we call it "objective" Bayesianism or O-Bayesianism, even though there is no agreement as to which of several systems is best. They all agree that there is no such thing as an "uninformative prior." Any attempted uninformative prior for a parameter becomes informative under transformations. Some strive to match the error probability (frequentist matching priors), but ironically they do not serve for the intended inference. It is one thing to argue the probability is .9999 that we would have rejected H_0^2—the so-called plain vanilla Higgs—if there was evidence for a particular type of BSM physics (and then argue because we have not rejected H_0^2, there is no such evidence). It is quite another to assign a .9999 posterior probability or degree of belief to H_0^2 itself. That clearly does not capture their epistemic stance about BSM. They expect violations; it is a matter of finding where they occur.

The interpretations of the posterior probabilities are unclear. The default priors are not regarded as beliefs; they may not even be probabilities, being improper (not summing to 1). In some treatments of the complex Higgs experiments, a host of default Bayesian estimates enter, but that is quite different from seeking a posterior probability in the Higgs, or the SM. "Even if there are some cases where good frequentist solutions are more neatly generated through Bayesian machinery, it would show only their technical value for goals that differ fundamentally from their own" (Cox and Mayo 2010, 301).

Then there is the problem of arriving at and testing the assumptions of the models themselves. Consider statistician George Box. Interestingly enough, given that he is at least half a Bayesian, Box shares our view of statistical induction, not as probabilism but as identifying and critically testing whether a model is "consonant" with particular facts. After arriving at an appropriate model, the move "is entirely *deductive* and will be called *estimation*" (Box 1983, 56). The deductive portion, he thinks, can be Bayesian, but the inductive portion requires frequentist significance tests, and statistical inference depends on an iteration between the two.

> Why can't all criticism be done using Bayes posterior analysis? . . . The difficulty with this approach is that by supposing all possible sets of assumptions are known a priori, it discredits the possibility of new discovery. But new discovery is, after all, the most important object of the scientific process. (Box 1983, 73).

The deepest problem, and the one of most interest for our purposes, is that a Bayesian model for experiment falls short of what is needed for discovery, for learning about novel and highly unexpected ways to extend a model. This is why many like George Box call for ecumenism.

Box considers imagining M_1, M_2, . . . , M_k as the alternative models and then computing $Pr(M_i|x)$, but he denies this is plausible. To assume we start out with all models precludes the "something else we haven't thought of" so vital to science (Box 1983, 73). Typically Bayesians try deal with this by computing a Bayesian catchall "everything else." The problem is, model M_i may get a high posterior probability relative to the other models considered, allowing one to overlook the one that actually gave rise to the data. Comparing posteriors of two models or reporting Bayes Factors have the same deficit: the more probable of the two does not point you to a model outside those considered. Further, Bayes factors are relative to the alternative one happens to choose, and its prior. The update from CERN since 2015 underscores these issues.

2016 Update

When the collider restarted in 2015, it had far greater collider energies than before. On December 15, 2015, something exciting happened:

> LHC's ATLAS and CMS experiments both reported a small "bump" in their data that denoted an excess of photon pairs with a combined mass of around 750 GeV. As this unexpected bump could be the first hint of a new massive particle that is not predicted by the Standard Model of particle physics, the data generated hundreds of theory papers that attempt to explain the signal.
>
> Taking into account what is known as the "look-elsewhere effect" (the fact that across a range of energies some bumps are bound to appear by chance), CMS says it has seen an excess with a statistical significance of 1.6σ, while ATLAS reports a significance of about 2σ—corresponding, respectively, to a roughly 1 in 10 and 1 in 20 chance that the result is a fluke.

> While these levels are far below the 5σ "gold standard" that must be met to claim a discovery, the fact that both collaborations saw a bump at the same energy has excited theoretical physicists. Indeed, since December, theorists have uploaded more than 250 papers on the subject to the *arXiv* preprint server. (Cartlidge 2016)[8]

> The significance reported by CMS is still far below physicists' threshold for a discovery: 5 sigma, or a chance of around 3 in 10 million that the signal is a statistical fluke. (Castelvecchi and Gibney 2016)

The new particle appears to have mass of 750 GeV (compared to 125 GeV for the Higgs).

> No theorist has ever predicted that such a particle should exist. No experiment has ever been designed to look for one. (*Nature* Editors 2016, 139)

> Physicists say that by June, or August at the latest, CMS and ATLAS should have enough data to either make a statistical fluctuation go away—if that's what the excess is—or confirm a discovery. (Castelvecchi and Gibney 2016)

Notice some highlights. First, the reports allude to the probability "the result is a fluke," and in the manner I have been proposing. As the data come in, either the significance levels will grow or wane as do the bumps. Physicists might object that it is the science writers who reword the physicists' careful U-type statements into D-type statements. There is evidence for that, but I think they are reacting to critical reports based on how things look from Bayesian probabilists' eyes.

Remember, for a Bayesian, once the data are known, they are fixed; what is random is an agent's beliefs or uncertainties on what is unknown—namely, the hypothesis. For the frequentist, considering the probability of $\{d(X) > d(x_0)\}$ is scarcely irrelevant even once $d(x_0)$ is known. It is the way to determine, following the severe testing principles, whether the null hypothesis can be falsified. ATLAS reports that on the basis of the P value display, "these results provide conclusive evidence for the discovery of a new particle with mass [approximately 125 GeV]" (Atlas Collaboration 2012b, 15).

Second, results are pointing to radically new physics that no one had thought of. It is hard to see how O'Hagan and Lindley's Bayesian model could do more than keep updating beliefs in the already expected parameters.

Perhaps eventually, one might argue, the catchall factor might become probable enough to start seeking something new, but it would be a far less efficient way for discovery than the simple significance tests.

The update illustrates our two-part analysis: first, is it real (and not a statistical fluke)? and second, if it is real, what are its properties? They do not want to assign a probability to its being genuine. If it disappears, it will be falsified, not made less probable. They will discover whether it is a statistical fluctuation of background, not the probability it is a statistical fluctuation. "Although these results are compatible with the hypothesis that the new particle is the Standard Model Higgs boson, more data are needed to assess its nature in detail" (Atlas Collaboration 2012b, 15).

CONCLUDING COMMENT

When a 5 sigma result in HEP is associated with claims such as "it's highly improbable our results are a statistical fluke" what is meant is that it is highly improbable to have been able to produce the display of bumps they do, with significance growing with more and better data, under the hypothesis of background alone. To turn the tables on the Bayesian, they may be illicitly sliding from what may be inferred from an entirely legitimate high probability. The reasoning is this: with probability .9999997, our methods would show that the bumps disappear, *under* the assumption data are due to background H_0. The bumps do not disappear but grow. We may infer a genuine effect. Using the same reasoning at the next stage they infer H^* is a Higgs-like particle. There have been misinterpretations of P values, for sure, but if a researcher has just described performing a statistical significance test, it would be ungenerous to twist probabilistic assertions into posterior probabilities. It would be a kind of "confirmation bias" whereby one insists on finding a sentence among very many that could be misinterpreted Bayesianly. The ASA report on P values contains many principles as to what P values are not. Principle 2 asserts:

> P-values do not measure [1] the probability that the studied hypothesis is true, or [2] the probability that the data were produced by random chance alone. (Wasserstein and Lazar 2016, 131)

I inserted the 1 and 2 absent from the original principle, because while 1 is true, phrases along the lines of 2 should not be equated to 1. I granted that claims along the lines of U-1 through U-3 are less open to misreading than their D-1 through D-3 counterparts. But Bayesian posteriors can also be misinterpreted as giving a frequentist performance, even though they do not.[9]

More productively, it should be seen that the 5 sigma report, or corresponding *P* value, is not the statistical inference, for a frequentist. Additional links along the lines of severe testing principles are needed to move from statistical information plus background (theoretical and empirical) to detach inferences. Having inferred *H**: a Higgs particle, one may say informally, "So probably we have experimentally demonstrated the Higgs," or "Probably, the Higgs exists." But whatever these might mean in informal parlance, they are not formal mathematical probabilities. Discussions on statistical philosophy must not confuse these.

Some might allege that I am encouraging a construal of *P* values that physicists have bent over backward to avoid. I admitted at the outset that "the problem is a bit delicate, and my solution is likely to be provocative." My position is simply that it is not legitimate to criticize frequentist measures from a perspective that assumes a very different role for probability. As Efron says, "The two philosophies represent competing visions of how science progresses" (2013, 130).

NOTES

1. Atlas Higgs experiment, public results. 2012. https://twiki.cern.ch/twiki/bin/view/AtlasPublic/HiggsPublicResults; CMS Higgs experiment, public results. 2012. https://twiki.cern.ch/twiki/bin/view/CMSPublic/PhysicsResultsHIG.

2. The Neyman–Pearson (NP) test recommends the test hypothesis be the one whose erroneous rejection would be the more serious.

3. The usual notation is $\{x: d(X) > .5\}$ and $\{x: Pr(X) < .3\}$.

4. The power of the a test to reject the null hypothesis at level α, when the true value of μ is equal to μ' is: $Pr(d(X) \geq d_\alpha; \mu = \mu')$.

5. The inference to (2) is a bit stronger than merely falsifying the null because certain properties of the particle must be shown at the second stage.

6. In searching for BSM science, hundreds of nulls have not been rejected, but no one considers this evidence that there is no BSM physics. If they find a nonstandard effect it is big news, but if they do not it is not grounds to reject BSM. Some claim this violates the Popperian testing principle of severity. It does not. The test only probes a given BSM physics; there

is not a high probability that all ways have been probed. This actually follows from the asymmetry of falsification and confirmation: all you need is an existential for an anomaly.

7. The original discussion, under the "Higgs boson" topic, on the International Society for Bayesian Analysis forum is archived here: https://web .archive.org/web/20160310031420/http://bayesian.org/forums/news/3648.

8. ArXiv.org 2016, search results at Cornell University Library for papers on the GeV 750 bump: http://arxiv.org/find/all/1/ti:+AND+750+Gev/0/1/0 /all/0/1.

9. For example, a posterior of 0.9 is often taken to mean the inference was arrived at by a method that is correct 90 percent of the time. This performance reading is not generally warranted.

REFERENCES

ATLAS Collaboration. 2012a. "Latest Results from ATLAS Higgs Search." Press statement. *ATLAS Updates,* July 4, 2012. http://atlas.cern/updates/press -statement/latest-results-atlas-higgs-search.

ATLAS Collaboration 2012b. "Observation of a New Particle in the Search for the Standard Model Higgs Boson with the ATLAS Detector at the LHC." *Physics Letters B* 716 (1): 1–29.

ATLAS Collaboration. 2012c. "Updated ATLAS Results on the Signal Strength of the Higgs-like Boson for Decays into *WW* and Heavy Fermion Final States." ATLAS-CONF-2012-162. *ATLAS Note,* November 14, 2012. http:// cds.cern.ch/record/1494183/files/ATLAS-CONF-2012-162.pdf.

Berger, James O. 2006. "The Case for Objective Bayesian Analysis" and "Rejoinder." *Bayesian Analysis* 1 (3): 385–402; 457–64.

Box, G. E. P. 1983. "An Apology for Ecumenism in Statistics." In *Scientific Inference, Data Analysis, and Robustness: Proceedings of a Conference Conducted by the Mathematics Research Center, the University of Wisconsin–Madison, November 6, 1981,* edited by G. E. P. Box, Tom Leonard, and Chien-Fu Wu, 51–84. New York: Academic Press.

Cartlidge, Edwin. 2016. "Theorizing about the LHC's 750 GeV Bump." *Physicsworld.com,* April 19, 2016. http://physicsworld.com/cws/article/news/2016 /apr/19/theorizing-about-the-lhcs-750-gev-bump.

Castelvecchi, Davide, and Elizabeth Gibney. 2016. "Hints of New LHC Particle Get Slightly Stronger." *Nature News,* March 19, 2016, doi:10.1038/nature .2016.19589.

CMS Experiment. 2012. "Observation of a New Particle with a Mass of 125 GeV." Press statement. *Compact Muon Solenoid Experiment at CERN* [blog], July 4, 2012. http://cms.web.cern.ch/news/observation-new-particle-mass-125-gev.

Cousins, Robert D. 2017. "The Jeffreys–Lindley Paradox and Discovery Criteria in High Energy Physics." *Synthese* 194 (2): 394–432.

Cox, David R., and D. V. Hinkley. 1974. *Theoretical Statistics*. London: Chapman and Hall.

Cox, David R., and Deborah Mayo. 2010. "Objectivity and Conditionality in Frequentist Inference." In *Error and Inference: Recent Exchanges on Experimental Reasoning, Reliability and the Objectivity and Rationality of Science*, edited by Deborah Mayo and Aris Spanos, 276–304. Cambridge: Cambridge University Press.

Efron, Bradley. 2013. "A 250-Year Argument: Belief, Behavior, and the Bootstrap." *Bulletin of the American Mathematical Society* 50: 129–46.

Fisher, Ronald. 1947. *The Design of Experiments*. 4th ed. Edinburgh: Oliver and Boyd.

Hacking, Ian. 1965. *Logic of Statistical Inference*. Cambridge: Cambridge University Press.

Hacking, Ian. 1980. "The Theory of Probable Inference: Neyman, Peirce and Braithwaite." In *Science, Belief and Behavior: Essays in Honour of R. B. Braithwaite*, edited by D. Mellor, 141–60. Cambridge: Cambridge University Press.

Ioannidis, John P. A. 2005. "Why Most Published Research Findings Are False." *PLoS Medicine* 2 (8): e124.

Lehmann, Erich L., and Joseph P. Romano. 2005. *Testing Statistical Hypotheses*. 3rd ed. New York: Springer.

Mayo, Deborah. 1996. *Error and the Growth of Experimental Knowledge*. Chicago: University of Chicago Press.

Mayo, Deborah. 2016. "Don't Throw Out the Error Control Baby with the Bad Statistics Bathwater: A Commentary." "The ASA's Statement on *p*-Values: Context, Process, and Purpose," *The American Statistician* 70 (2) (supplemental materials).

Mayo, Deborah. 2018. *Statistical Inference as Severe Testing: How to Get Beyond the Statistics Wars*. Cambridge University Press.

Mayo, Deborah, and David R. Cox. 2006. "Frequentist Statistics as a Theory of Inductive Inference." In *Optimality: The Second Erich L. Lehmann Symposium*, edited by J. Rojo. Lecture Notes-Monograph series, Institute of Mathematical Statistics (IMS) 49: 77-97.

Mayo, Deborah, and Aris Spanos. 2006. "Severe Testing as a Basic Concept in a Neyman–Pearson Philosophy of Induction." *British Journal for the Philosophy of Science* 57: 323–57.

Mayo, Deborah, and Aris Spanos. 2011. "Error Statistics." In *Philosophy of Statistics,* vol. 7 of *Handbook of the Philosophy of Science,* edited by Dov M. Gabbay, Prasanta S. Bandyopadhyay, Malcolm R. Forster, Paul Thagard, and John Woods, 153–98. Amsterdam: Elsevier.

Nature Editors. 2016. "Who Ordered That?" *Nature* 531: 139–40. doi:10.1038/531139b.

O'Hagan, Tony. 2012. "Higgs Boson—Digest and Discussion." August 20, 2012. https://web.archive.org/web/20150508154446/http://tonyohagan.co.uk/academic/pdf/HiggsBoson.pdf.

Overbye, Dennis. 2013. "Chasing the Higgs Boson." *New York Times,* March 15, 2013. http://www.nytimes.com/2013/03/05/science/chasing-the-higgs-boson-how-2-teams-of-rivals-at-CERN-searched-for-physics-most-elusive-particle.html.

Peirce, Charles S. 1931–35. *Collected Papers, Volumes 1–6.* Edited by C. Hartsthorne and P. Weiss. Cambridge: Harvard University Press.

Reich, Eugenie Samuel. 2012. "Flaws Found in Faster Than Light Neutrino Measurement." *Nature News,* February 22, 2012. doi:10.1038/nature.2012.10099.

Spiegelhalter, David. 2012. "Explaining 5-Sigma for the Higgs: How Well Did They Do?" *Understanding Uncertainty* [blog], August 7, 2012. https://understandinguncertainty.org/explaining-5-sigma-higgs-how-well-did-they-do.

Staley, Kent W. 2017. "Pragmatic Warrant for Frequentist Statistical Practice: The Case of High Energy Physics." *Synthese* 194 (2): 355–76.

Strassler, Matt. 2013a. "CMS Sees No Excess in Higgs Decays to Photons." *Of Particular Significance* [blog], March 12, 2013. https://profmattstrassler.com/2013/03/14/cms-sees-no-excess-in-higgs-decays-to-photons/.

Strassler, Matt. 2013b. "A Second Higgs Particle?" *Of Particular Significance* [blog], July 2, 2013. https://profmattstrassler.com/2013/07/02/a-second-higgs-particle/.

Wasserman, Larry. 2012. "The Higgs Boson and the p-Value Police." *Normal Deviate* [blog], July 11, 2012. http://normaldeviate.wordpress.com/2012/07/11/the-higgs-boson-and-the-p-value-police/.

Wasserstein, Ronald L., and Nicole A Lazar. 2016. "The ASA's Statement on p-Values: Context, Process and Purpose." *American Statistician* 70: 129–33.

8

VALUES AND EVIDENCE IN MODEL-BASED
CLIMATE FORECASTING

ERIC WINSBERG

No CONTEMPORARY VOLUME on the experimental side of modeling in science would be complete without a discussion of the "model experiments" being run in contemporary climate science. I am referring here, of course, to the enormous, complex, and elaborate global climate simulations being run to try to predict the pace and tempo of climate change, both at the global and regional level, over the course of the next century.[1] And no section on evidential reasoning in science would be complete without some discussion of the appropriate role, if there is any, of social and ethical values in reasoning about scientific models, hypotheses, and predictions.

Fortunately, climate science and the computer simulation of climate in particular provide an excellent background against which to discuss the role of values in science. One reason for this is obvious: climate science is one of the most policy-driven disciplines in the scientific world. Outside of the health sciences, no area of science is more frequently called on to inform public policy and public action, and is thus more frequently wedded to aspects of public decision making that themselves draw on social and ethical value judgments. But there is a second, less obvious reason that climate science is an excellent place to study the role of values in the physical sciences. This has to do with a now-famous and widely accepted defense of the value-free nature of science that was first put forward in the 1950s by Richard Jeffrey. We will see more details in the sequel, but his basic idea was that the key to keeping science free of social and ethical values—and hence in preserving its objectivity—was for scientists to embrace probabilistic reasoning and to attach estimations of uncertainties to all their claims to knowledge.

But it is precisely this endeavor—the attempt to attach estimations of the degree of uncertainty to the predictions of global climate models—that has raised in the domain of climate science some of the most persistent and troubling conceptual difficulties. There has been over the last several years an explosion of interest and attention devoted to this problem—the problem of uncertainty quantification (UQ) in climate science. The technical challenges associated with this project are formidable: the real data sets against which model runs are evaluated are large, patchy, and involve a healthy mixture of direct and proxy data; the computational models themselves are enormous, and hence the number of model instances that can be run is minuscule and sparsely distributed in the solution space that needs to be explored; the parameter space that we would like to sample is vast and multidimensional; and the structural variation that exists among the existing set of models is substantial but poorly understood. Understandably, therefore, the statistical community that has engaged itself with this project has devoted itself primarily to overcoming some of these technical challenges. But as these technical challenges are being met, a number of persistent conceptual difficulties remain. What lessons can we draw from these conceptual difficulties concerning the proper role of social and ethical values in science?

SCIENCE AND SOCIAL VALUES

What do we mean, first of all, by social values? Social values, I take it, are the estimations of any agent or group of agents of what is important and valuable—in the typical social and ethical senses—and what is to be avoided, and to what degree. What value does one assign to economic growth compared with the degree to which we would like to avoid various environmental risks? In the language of decision theory, by social values we mean the various marginal utilities one assigns to events and outcomes. The point of the word "social" in "social values" is primarily to flag the difference between these values and what Ernan McMullin (1983) once called "epistemic values," such as simplicity and fruitfulness. But I do not want to beg any questions about whether values that are paradigmatically ethical or social can or cannot, or should or should not, play important epistemic roles. So I prefer not to use that vocabulary. I talk instead about social and ethical values when I am referring to things that are valued for paradigmatically social or ethical reasons. I do not carefully distinguish in this discussion between the social and the ethical.[2]

The philosophically controversial question about social and ethical values is about the degree to which they are involved (or better put: the degree to which they are necessarily involved, or inevitably involved, and perhaps most importantly uncorrectably involved) in the appraisal of hypotheses, theories, models, and predictions. This is the question, after all, of the degree to which the epistemic and the normative can be kept apart.

This is a question of some importance because we would like to believe that only experts should have a say in what we ought to believe about the natural world. But we also think that it is *not* experts, or at least not experts *qua* experts, who should get to say what is important to us, or what is valuable, or has utility. Such a division of labor, however, is only possible to the extent that the appraisal of scientific hypotheses, and other matters that require scientific expertise, can be carried out in a manner that is free of the influence of social and ethical values.

Philosophers of science of various stripes have mounted a variety of arguments to the effect that the epistemic matter of appraising scientific claims of various kinds cannot be kept free of social and ethical values.[3] Here, we will be concerned only with one such line of argument—one that is closely connected to the issue of UQ—that goes back to the midcentury work of the statistician C. West Churchman (1948, 1956) and a philosopher of science, Richard Rudner (1953). This line of argument is now frequently referred to as the argument from *inductive risk*. It was first articulated by Rudner in the following schematic form:

1. The scientist qua scientist accepts or rejects hypotheses.
2. No scientific hypothesis is ever completely (with 100% certainty) verified.
3. The decision to either accept or reject a hypothesis depends upon whether the evidence is sufficiently strong.
4. Whether the evidence is *sufficiently* strong is "a function of the *importance*, in a typically ethical sense, of making a mistake in accepting or rejecting the hypothesis."
5. Therefore, the scientist qua scientist makes value judgments.

Rudner's oft repeated example was this: How sure do you have to be about a hypothesis if it says (1) a toxic ingredient of a drug is not present in lethal quantity, versus (2) a certain lot of machine-stamped belt buckles is not defective. "How sure we need to be before we accept a hypothesis will depend upon how serious a mistake would be?" (Rudner 1953, 2). We can easily translate Rudner's lesson into an example from climate science. Should we accept or reject the hypothesis, for example, that, given future emissions trends, a certain regional climate outcome will occur? Should we accept the hypothesis, let us say, that a particular glacial lake dam will burst in the next fifty years? Suppose that, if we accept the hypothesis, we will replace the moraine with a concrete dam. But whether we want to build the dam will depend not only on our degree of evidence for the hypothesis, but also on how we would measure the severity of the consequences of building the dam and having the glacier not melt, versus not building the dam and having the glacier melt. Thus, Rudner would have us conclude that so long as the evidence is not 100 percent conclusive we cannot justifiably accept or reject the hypothesis without making reference to our social and ethical values.

The best known reply to Rudner's argument came from the philosopher, logician, and decision theorist Richard Jeffrey (1956). Jeffrey argued that the first premise of Rudner's argument, that it is the proper role of the scientist qua scientist to accept and reject hypotheses, is false. Their proper role, he urged, is to assign probabilities to hypotheses with respect to the currently available evidence. Others, for example policy makers, can attach values or utilities to various possible outcomes or states of affairs and, in conjunction with the probabilities provided by scientists, decide how to act.

It is clear that Jeffrey did not anticipate the difficulties that modern climate science would have with the task that he expected to be straightforward and value free, the assignment of probability with respect to the available evidence. There are perhaps many differences between the kinds of examples that Rudner and Jeffrey had in mind and the kinds of situations faced by climate scientists. For one, Rudner and Jeffrey discuss cases in which we need the probability of the truth or falsity of a single hypothesis, but climate scientists generally are faced with having to assign probability distributions over a space of possible outcomes.

I believe that the most significant difference between the classic kind of inductive reasoning Jeffrey had in mind (in which the probabilities scientists are meant to offer are their subjective degrees of belief based on the available evidence) and the contemporary situation in climate science is the extent to which epistemic agency in climate science is distributed across a wide range of scientists and tools. This is a point that will receive more attention later in this discussion. For now, we should turn to what I would claim are typical efforts in climate science to deliver probabilistic forecasts, and we will see how they fare in terms of Jeffrey's goal to use probabilities to divide labor between the epistemic and the normative.

UNCERTAINTY IN CLIMATE SCIENCE

Where do probabilistic forecasts in climate science come from? We should begin with a discussion of the sources of uncertainty in climate models. There are two main sources that concern us here: *structural model uncertainty* and *parameter uncertainty*. Although the construction of climate models is guided by basic science—science in which we have a great deal of confidence—these models also incorporate a barrage of auxiliary assumptions, approximations, and parameterizations, all of which contribute to a degree of uncertainty about the predictions of these models. Different climate models (with different basic structures) produce substantially different predictions. This source of uncertainty is often called *structural model uncertainty.*

Next, complex models involve large sets of *parameters* or aspects of the model that have to be quantified before the model can be used to run a simulation of a climate system. We are often highly uncertain about what the best value for many of these parameters is, and hence, even if we had at our disposal a model with ideal (or perfect) structure, we would still be uncertain about the behavior of the real system we are modeling because the same model structure will make different predictions for different values of the parameters. Uncertainty from this source is called *parameter uncertainty.*[4]

Most efforts in contemporary climate science to measure these two sources of uncertainty focus on what one might call *sampling methods.* In practice, in large part because of the high computational cost of each model run, these methods are extremely technically sophisticated, but in principle they are rather straightforward.

I can best illustrate the idea of sampling methods with an example regarding *parameter uncertainty.* Consider a simulation model with one parameter and several variables.[5] If one has a data set against which to benchmark the model, one could assign a weighted score to each value of the parameter based on how well it retrodicts values of the variables in the available data set. Based on this score, one could then assign a probability to each value of the parameter. Crudely speaking, what we are doing in an example like this is observing the frequency with which each value of the parameter is successful in replicating known data. How many of the variables does it get right? With how much accuracy? Over what portion of the time history of the data set? And then we weight the probability of the parameter taking this value in our distribution in proportion to how well it had fared in those tests.

The case of structural model uncertainty is similar. The most common method of estimating the degree of structural uncertainties in the predictions of climate models is a set of sampling methods called "ensemble methods." The core idea is to examine the degree of variation in the predictions of the existing set of climate models that happen to be on the market. By looking at the average prediction of the set of models and calculating their standard deviation, one can produce a probability distribution for every value that the models calculate.

SOME WORRIES ABOUT THE STANDARD METHODS

There are reasons to doubt that these kinds of straightforward methods for estimating both structural model uncertainty and parameter uncertainty are conceptually coherent. Signs of this are visible in the results that have been produced. These signs have been particularly well noted by the climate scientists Claudia Tebaldi and Reto Knutti (2007), who note, in the first instance, that many studies founded on the same basic principles produce radically different probability distributions.

Indeed, I would argue that there are four reasons to suspect that ensemble methods are not a conceptually coherent set of methods.

1. Ensemble methods either assume that all models are equally good, or they assume that the set of available methods can be relatively weighted.
2. Ensemble methods assume that, in some relevant respect, the set of available models represent something like a sample of independent draws from the space of possible model structures.
3. Climate models have shared histories that are very hard to sort out.
4. Climate modelers have a herd mentality about success.

I will discuss each of these four reasons in what follows. But first, consider a simple example that mirrors all four elements on the list. Suppose that you would like to know the length of a barn. You have one tape measure and many carpenters. You decide that the best way to estimate the length of the barn is to send each carpenter out to measure the length, and you take the average. There are four problems with this strategy. First, it assumes that each carpenter is equally good at measuring. But what if some of the carpenters have been drinking on the job? Perhaps you could weight the degree to which their measurements play a role in the average in inverse proportion to how much they have had to drink. But what if, in addition to drinking, some have also been sniffing from the fuel tank? How do you weight these relative influences? Second, you are assuming that each carpenter's measurement is independently scattered around the real value. But why think this? What if there is a systematic error in their measurements? Perhaps there is something wrong with the tape measure that systematically distorts them. Third (and relatedly), what if all the carpenters went to the same carpentry school, and they were all taught the same faulty method for what to do when the barn is longer than the tape measure? And fourth, what if each carpenter, before they record their value, looks at the running average of the previous measurements, and if theirs deviates too much they tweak it to keep from getting the reputation as a poor measurer?

All these sorts of problems play a significant role—both individually and, especially, jointly—in making ensemble statistical methods in climate science conceptually troubled. I will now discuss the role of each of them in climate science in detail.

1. Ensemble methods either assume that all models are equally good, or they assume that the set of available methods can be relatively weighted.

If you are going to use an ensemble of climate models to produce a probability distribution, you ought to have some grounds for believing that all of them ought to be given equal weight in the ensemble. Failing that, you ought to have some principled way to weight them. But no such thing seems to exist. Although there is widespread agreement among climate scientists that some models are better than others, quantifying this intuition seems to be particularly difficult. It is not difficult to see why.

As Gleckler, Taylor, and Doutriaux (2008) point out, no single metric of success is likely to be useful for all applications. They carefully studied the success of various models at various prediction tasks. They showed that there are some unambiguous flops on the list and no unambiguous winner—and no clear way to rank them.

2. Ensemble methods assume that, in some relevant respect, the set of available models represent something like a sample of independent draws from the space of possible model structures.

This is surely the greatest problem with ensemble statistical methods. The average and standard deviation of a set of trials is only meaningful if those trials represent a random sample of independent draws from the relevant space—in this case, the space of possible model structures. Many commentators have noted that this assumption is not met by the set of climate models on the market. In fact, I would argue it is not exactly clear what this would even mean in this case. What, after all, *is* the space of possible model structures? And why would we want to sample randomly from this? After all, we want our models to be as physically realistic as possible, not random. Perhaps we are meant to assume instead that the existing models are randomly distributed around the ideal model, in some kind of normal distribution. This would be an analogy to measurement theory. But modeling is not measurement, and there is very little reason to think this assumption holds.[6]

3. Climate models have shared histories that are very hard to sort out.

One obvious reason to doubt that the last assumption is valid is that large clusters of the climate models on the market have shared histories.[7] Some models share code, many scientists move from one laboratory to another and bring ideas with them, some parts of climate models (though not physically principled) are from a common toolbox of techniques, and so on. Worse still, we do not even have a systematic understanding of these inter-relations. So it is not just the fact that most current statistical ensemble methods are naïve with respect to these effects, but that it is far from obvious that we have the background knowledge we would need to eliminate this "naïveté"—to account for them statistically.

4. Climate modelers have a herd mentality about success.

Herd mentality is a frequently noted feature of climate modeling. Most climate models are highly tunable with respect to some of their variables, and to the extent that no climate laboratory wants to be the oddball on the block there is significant pressure to tune one's model to the crowd. This kind of phenomenon has historical precedent. In 1939 Walter Shewhart published a chart of the history of measurement of the speed of light. The chart showed a steady convergence of measured values that was not well explained by their actual success. Myles Allen (2008) put the point like this: "If modelling groups, either consciously or by 'natural selection,' are tuning their flagship models to fit the same observations, spread of predictions becomes meaningless: eventually they will all converge to a delta-function."

THE INEVITABILITY OF VALUES: DOUGLAS *CONTRA* JEFFREY

What should we make of all these problems from the point of view of the Rudner–Jeffrey debate? This much should be clear: from the point of view of Jeffrey's goal, to separate the epistemic from the normative, UQ based on statistical ensemble methods will not do. But this much should have been clear from Heather Douglas's (2000) discussion of the debate about science and values.[8]

Douglas noted a flaw in Jeffrey's response to Rudner. She remarked that scientists often have to make methodological choices that do not lie on a continuum. Douglas points out that which choice I make will depend on my inductive risk profile. To the extent that I weigh more heavily the consequences of saying that the hypothesis is false if it is in fact true, I will choose a method with a higher likelihood of false positives. And vice versa. But that, she points out, depends on my social and ethical values. Social and ethical values, she concludes, play an inevitable role in science.

There are at least two ways in which methodological choices in the construction of climate models will often ineliminably reflect value judgments in the typically social or ethical sense.

1. Model choices have reflected balances of inductive risk.
2. Models have been optimized, over their history, to particular purposes and to particular metrics of success.

The first point should be obvious from our discussion of Rudner. When a climate modeler is confronted with a choice between two ways of solving a modeling problem, she may be aware that each choice strikes a different balance of inductive risks with respect to a problem that concerns her at the time. Choosing which way to go in such a circumstance will have to reflect a value judgment. This will always be true so long as a methodological choice between methods A and B are not epistemologically *forced* in the following sense: while option A can be justified on the grounds that it is *less* likely to predict, say, outcome O than B is when O *will not* in fact occur, option B could also be preferred on the grounds that it is *more* likely to predict O if O *will* in fact occur.

The second point is that when a modeler is confronted with a methodological choice she will have to decide which metric of success to use when evaluating the success of the various possibilities. And it is hard to see how choosing a metric of success will not reflect a social or ethical value judgment or possibly even a response to a political pressure about which prediction task is more "important" (in a not-purely-epistemic sense). Suppose choice A makes a model that looks better at matching existing precipitation data, but choice B is better at matching temperature data. She will need to decide which prediction task is more important in order to decide which method of evaluation to use, and that will influence which methodological choice is pursued.

I think this discussion should make two things clear. First, ensemble sampling approaches to UQ are founded on conceptually shaky ground. Second, and perhaps more importantly, they do not enable UQ to fulfill its primary function—to divide the epistemic from the normative in the way that Jeffrey expected probabilistic forecasts to do. And they fail for just the reasons that Douglas made famous: because they ossify past methodological choices (which themselves can reflect balances of inductive risk and other social and ethical values) into "objective" probabilistic facts.

This raises, of course, the possibility that climate UQ could respond to these challenges by avoiding the use of "objective" statistical ensemble methods and by adopting more self-consciously Bayesian methods that attempt to elicit the expert judgment of climate modelers about their subjective degrees of belief concerning future climate outcomes. Call this the Bayesian response to the Douglas challenge (BRDC).

Indeed, this approach to UQ has been endorsed by several commenters on the problem.[9] Unfortunately, the role of genuinely subjective Bayesian approaches to climate UQ has been primarily in theoretical discussions of what to do, rather than in actual estimates that one sees published and that are delivered to policy makers. Here, I would to point to some of the difficulties that might explain this scarcity. Genuinely Bayesian approaches to UQ in climate science, in which the probabilities delivered reflect the expert judgment of climate scientists rather than observed frequencies of model outputs, face several difficulties. In particular, they arise as a consequence of three features of climate models: their massive *size and complexity;* the extent to which *epistemic agency* in climate modeling *is distributed,* in both time and space, across a wide range of individuals; and the degree to which *methodological choices* in climate models *are generatively entrenched.* I will try to say a bit about what I mean by each of these features in the next section.

THREE FEATURES OF CLIMATE MODELS

Size and Complexity

Climate models are enormous and complex. Take one of the state of the art American climate models, the U.S. National Oceanic and Atmospheric Administration (NOAA) Geophysical Fluid Dynamics Laboratory (GFDL) CM2.x. The computational model itself contains over a million lines of code. There are over a thousand different parameter options. It is said to involve modules that are "constantly changing" and involve hundreds of initialization files that contain "incomplete documentation." The CM2.x is said to contain novel component modules written by over a hundred different people. Just loading the input data into a simulation run takes over two hours. Using over a hundred processors running in parallel, it takes weeks to produce one model run out to the year 2100, and months to reproduce thousands of years of paleoclimate.[10] If you store the data from a state-of-the-art general circulation model (GCM) every five minutes, they can produce tens of terabytes per model year.

Another aspect of the models' complexity is their extreme "fuzzy modularity" (Lenhard and Winsberg, 2010). In general, a modern state-of-the-art climate model has a theoretical core that is surrounded and supplemented by various submodels that themselves have grown into complex entities. The interaction of all of them determines the dynamics. And these interactions are themselves quite complex. The coupling of atmospheric and oceanic circulation models, for example, is recognized as one of the milestones of climate modeling (leading to so-called coupled GCMs). Both components have had their independent modeling history, including an independent calibration of their respective model performance. Putting them together was a difficult task because the two submodels now interfered dynamically with each other.[11]

Today, atmospheric GCMs have lost their central place and given way to a deliberately modular architecture of coupled models that comprise a number of highly interactive submodels, such as atmosphere, oceans, or ice cover. In this architecture the single models act (ideally!) as interchangeable modules.[12] This marks a turn from one physical core—the fundamental equations of atmospheric circulation dynamics—to a more networked picture of interacting models from different disciplines (Küppers and Lenhard 2006).

In sum, climate models are made up of a variety of modules and submodels. There is a module for the general circulation of the atmosphere, a module for cloud formation, for the dynamics of sea and land ice, for effects of vegetation, and many more. Each of them, in turn, includes a mixture of principled science and parameterizations. And it is the interaction of these components that brings about the overall observable dynamics in simulation runs. The results of these modules are not first gathered independently and only after that synthesized; rather, data are continuously exchanged between all modules during the runtime of the simulation.[13] The overall dynamics of one global climate model are the complex result of the interaction of the modules—not the interaction of the results of the modules. This is why I modify the word "modularity" with the warning flag "fuzzy" when I talk about the modularity of climate models: due to interactivity and the phenomenon of "balance of approximations," modularity does not break down a complex system into separately manageable pieces.[14]

Distributed Epistemic Agency

Climate models reflect the work of hundreds of researchers working in different physical locations and at different times. They combine incredibly diverse kinds of expertise, including climatology, meteorology, atmospheric dynamics, atmospheric physics, atmospheric chemistry, solar physics, historical climatology, geophysics, geochemistry, geology, soil science, oceanography, glaciology, paleoclimatology, ecology, biogeography, biochemistry, computer science, mathematical and numerical modeling, time series analysis, and others.

Not only is epistemic agency in climate science distributed across space (the science behind model modules comes from a variety of laboratories around the world) and domains of expertise, but also across time. No state-of-the-art, coupled atmosphere-ocean GCM (AOGCM) is literally built from the ground up in one short surveyable unit of time. They are assemblages of methods, modules, parameterization schemes, initial data packages, bits of code, and coupling schemes that have been built, tested, evaluated, and credentialed over years or even decades of work by climate scientists, mathematicians, and computer scientists of all stripes.[15]

No single person—indeed no group of people in any one place, at one time, or from any one field of expertise—is in any position to speak authoritatively about any AOGCM in its entirety.

Methodological Choices Are Generatively Entrenched

Johannes Lenhard and I (2010) have argued that complex climate models acquire an intrinsically historical character and show path dependency. The choices that modelers and programmers make at one time about how to solve particular problems of implementation have effects on what options will be available for solving problems that arise at a later time. And they will have effects on what strategies will succeed and fail. This feature of climate models, indeed, has led climate scientists such as Smith (2002) and Palmer (2001) to articulate the worry that differences between models are concealed in code that cannot be closely investigated in practice. We called this feature of climate models *generative entrenchment* and argued that it leads to an analytical impenetrability of climate models; we have been unable—and are likely to continue to be unable—to attribute all or perhaps even most of the various sources of their successes and failures to their internal modeling assumptions.

This last claim should be clarified to avoid misunderstanding. As we have seen, different models perform better under certain conditions than others. But if model A performs better at making predictions on condition A*, and model B performs better under condition B*, then optimistically one might hope that a hybrid model—one that contained some features of model A and some features of model B—would perform well under both sets of conditions. But what would such a hybrid model look like?

Ideally, to answer that question one would like to attribute the success of each of the models A and B to the success of their particular submodels—or components. One might hope, for example, that a GCM that is particularly good at predicting precipitation is one that has, in some suitably generalizable sense, a particularly good rain module. We called success in such an endeavor, the process of teasing apart the sources of success and failure of a simulation, "analytic understanding" of a global model. We would say that one has such understanding precisely when one is able to identify the extent to which each of the submodels of a global model is contributing to its various successes and failures.

Unfortunately, analytic understanding is extremely hard to achieve in this context. The complexity of interaction between the modules of the simulation is so severe, as is the degree to which balances of approximation play an important role, that it becomes impossible to independently assess the merits or shortcomings of each submodel. One cannot trace back the effects of assumptions because the tracks get covered during the kludging together of complex interactions. This is what we called "analytic impenetrability" (Lenhard and Winsberg, 2010, 261). But analytic impenetrability makes epistemically inscrutable the effects on the success and failure of a global model of the past methodological assumptions that are generatively entrenched.

Summary

State-of-the-art global climate models are highly complex, they are the result of massively distributed epistemic labors, and they arise from a long chain of generatively entrenched methodological choices whose effects are epistemically inscrutable. These three features, I would now argue, make the BRDC very difficult to pull off with respect to climate science.

THE FAILURE OF THE BRDC IN CLIMATE SCIENCE

Recall how the BRDC is meant to go. Rudner argues that the scientist who accepts or rejects hypotheses has to make value judgments. Jeffrey replies that she should only assign probabilities to hypotheses on the basis of the available evidence and, in so doing, avoid making value judgments. Douglas argues that scientists make methodological choices and that these choices will become embedded in the mix of elements that give rise to estimates of probabilities that come from classical, as opposed to Bayesian, statistics. Because those methodological choices will involve a balance of inductive risks, the scientist cannot avoid value judgments. The BRDC suggests that scientists avoid employing any deterministic algorithm that will transmit methodological choices into probabilities (such as employing a classical statistical hypothesis test in the toxicology case, or employing ensemble averages in the climate case) and should instead rely on their expert judgment to assess what the appropriate degree of belief in a hypothesis is, given that a particular methodological choice is made and resultant evidence acquired. The probabilities such a scientist would offer should be the scientist's subjective degree of belief, one that has been conditionalized on the available evidence.

There are, in fact, other methods for estimating probabilities in climate science that lean more heavily on the subjective judgment of experts, such as expert elicitation methods in which priors and likelihoods are obtained by asking scientists for their degrees of belief in various hypotheses. Imagining that these estimates are free from the values I have discussed overlooks the fact that the judgments of experts can be shaped significantly by their experience working with particular models and by their awareness of particular modeling results. Indeed, models are used in the study of complex systems in part *because* it can be very difficult to reason (accurately) about such systems without them. For sciences in which complex, nonlinear models play a prominent role, it is naïve to think that scientists can typically reach conclusions about probabilities that are both adequately well-informed *and* independent of the set of models that happen to be available to them at any particular time.

VALUES IN THE NOOKS AND CRANNIES

At this point in the discussion, it might be natural for a reader to ask for a specific example of a social, political, or ethical value that has influenced a methodological choice in the history of climate modeling. It is easy to give a couple of potted examples. In previous work, I have focused on what I have here labeled the second kind of role of values: that climate models have been optimized, over their history, to particular purposes and to particular metrics of success.[16] I gave the example that in the past modelers had perhaps focused on the metric of successfully reproducing known data about global mean surface temperature, rather than other possible metrics. And I speculated that they might have done so because of a social and political climate in which the concern was about "global warming" rather than the now more popular phrase "anthropogenic climate change."

But I now think it was a mistake to focus on particular historical claims about specific motives and choices. I want to focus instead on the fact that climate modeling involves literally thousands of unforced methodological choices. Many crucial processes are poorly understood, and many compromises in the name of computational exigency need to be made. All one needs to see is that, as in the case of the biopsy stain in the toxicology case, no unforced methodological choice can be defended in a value vacuum. If one asks, "Why parameterize this process rather than try to resolve it on the grid?" or "Why use this method for modeling cloud formation?" it will rarely be the case that the answer can be "because that choice is objectively better than the alternative." As some of the examples I have provided clearly illustrate, most choices will be better in some respects and worse in other respects than their alternatives, and the preference for the one over the other will reflect the judgment that one aspect is more important. Some choices will invariably increase the probability of finding a certain degree of climate variation, while its alternative will do the opposite—and so the choice that is made can be seen as reflecting a balance of inductive risks.

Kevin Elliot (2011b, 55) has identified three conditions under which scientists should be expected to incorporate social and ethical values in particular scientific cases: (1) the "ethics" principle, that scientists have an ethical responsibility to consider the impact of their methodological choices on society in the case under consideration; (2) the "uncertainty" principle, that the available scientific information is uncertain or incomplete; and (3) the "no-passing-the-buck" principle, that scientists cannot just withhold their judgment or give value-free information to policy makers and let them deal with the social/ethical issues. That the second condition is in play in climate science is clear. That the third one is in play follows from the failure of the BRDC.

How do we know that the first one is in play without mention of particular historical claims about specific motives and choices? I think all we need to argue here is that many of the choices made by climate modelers had to have been unforced in the absence of a relevant set of values—that in retrospect such choices could only be defended against *some set of predictive preferences* and *some balance of inductive risks*. In other words, any rational reconstruction of the history of climate science would have to make mention of predictive preferences and inductive risks at pain of making most of these choices seem arbitrary. But what I want to be perfectly clear about here (in a way that I think I have not been in earlier work) is that I do not mean to attribute to the relevant actors these psychological motives, nor any particular specifiable or recoverable set of interests.[17] I am not in the business of making historical, sociological, or psychological claims. I have no idea why individual agents made the choices that they made—and indeed it is part of my argument that these facts are mostly hidden from view. In fact, for many of the same reasons that these methodological choices are immune from the BRDC, they are also relatively opaque to us from a historical, philosophical, and sociological point of view. They are buried in the historical past under the complexity, epistemic distributiveness, and generative entrenchment of climate models.

Some readers may find that this makes my claim about the value-ladenness of climate models insufficiently concrete to have any genuine bite. One might ask, "Where are the actual values?" Some readers, in other words, might be craving some details about how agents have been specifically motivated by genuine concrete ethical or political considerations. They might be tempted to think that I have too abstractly identified the role of values here to be helpful. But this is to miss the dialectical structure of my point. The very features that make the BRDC implausible make this demand unsatisfiable. No help of the sort that "finds the hidden values" can be forthcoming on my account. The social, political, and ethical values that find their way into climate models cannot be recovered in bite-sized pieces.

Recall that we began this whole discussion with a desire to separate the epistemic from the normative. But we have now learned that, with respect to science that relies on models that are sufficiently complex, epistemically distributed, and generatively entrenched, it becomes increasingly difficult to tell a story that maintains that kind of distinction. And without being able to provide a history that respects that distinction, there is no way to isolate the values that have been involved in the history of climate science.

One consequence of the blurred distinction between the epistemic and the normative in our case is that the usual remarks philosophers often make about the value-ladenness of science do not apply here. Those who make the claim that science is value-laden often follow up with the advice that scientists ought to be more self-conscious in their value choices, and that they ought to ensure that their values reflect those of the people they serve. Or they suggest implementing some system for soliciting public opinions or determining public values and making that the basis for these determinations. But in the picture I am painting neither of these options is really possible. The bits of value-ladenness lie in all the nooks and crannies, might very well have been opaque to the actors who put them there, and are certainly opaque to those who stand at the end of the long, distributed, and path-dependent process of model construction. In the case of the biopsy stains I can say, "Consumer protection is always more important than corporate profits! Even in the absence of epistemologically forcing consideration, the toxicologist should choose the stain on the left!" But in the climate case, the situation is quite different. We can of course ask for a climate science that does not reflect systematic biases, like one that is cynically paid for by the oil industry. But this kind of demand for a science that reflects the "right values" cannot go all the way down into all those nooks and crannies. In those relevant respects, it becomes terribly hard to ask for a climate science that reflects "better" values.

NOTES

1. I would like to emphasize that the focus of this chapter is on that topic: attempts to predict the *pace and tempo* of future climate change, rather than on the question of whether climate change is happening and whether humans are its cause. The first topic is a topic of scientific controversy with interesting epistemological issues. The second topic is not.

2. And when I variously use the expressions "social values," "ethical values," or "social and ethical values," these differences in language should not be read as flagging important philosophical differences.

3. In addition to Churchman and Rudner, see also Frank (1954), Neurath (1913/1983), Douglas (2000), Howard (2006), Longino (1990, 1996, 2002), Kourany (2003a, 2003b), Solomon (2001), Wilholt (2009), and Elliott (2011a, 2011b).

4. Many discussions of UQ in climate science will also identify *data uncertainty*. In evaluating a particular climate model, including both its structure and parameters, we compare the model's output to real data. Climate modelers, for example, often compare the outputs of their models to records of past climate. These records can come from actual meteorological observations or from proxy data—inferences about past climate drawn from such sources as tree rings and ice core samples. Both of these sources of data, however, are prone to error, so we are uncertain about the precise nature of the past climate. This, in turn, has consequences for our knowledge of the future climate. Although data uncertainty is a significant source of uncertainty in climate modeling, I will not discuss this source of uncertainty here. For the purposes of this discussion, I make the crude assumption that the data against which climate models are evaluated are known with certainty. Notice, in any case, that data uncertainty is part of parameter uncertainty and structural uncertainty because it acts by affecting our ability to judge the accuracy of our parameters and our model structures.

5. A parameter for a model is an input that is fixed for all time whereas a variable takes a value that varies with time. A variable for a model is thus both an input for the model (the value the variable takes at some initial time) and an output (the value the variable takes at all subsequent times). A parameter is simply an input.

6. Some might argue that if we look at how the models perform on past data (for, say, mean global surface temperature), they often are distributed around the observations. But, first, these distributions do not display anything

like random characteristics (i.e., normal distribution). And, second, this feature of one variable for past data (the data for which the models have been tuned) is a poor indicator that it might obtain for all variables and for future data.

7. An article by Masson and Knutti, "Climate Model Geneology" (2011), discusses this phenomenon and its effects on multimodel sampling in detail.

8. Which, inter alia, did much to bring the issue of "inductive risk" back into focus for contemporary philosophy of science and epistemology.

9. See, for example, Goldstein and Rougier (2006).

10. All the above claims and quotations come from Dunne (2006), communicated to the author by John Dunne and V. Balaji, in personal correspondence.

11. For an account of the controversies around early coupling, see Shackley et al. (1999); for a brief history of modeling advances, see Weart (2010).

12. As, for example, in the Earth System Modeling Framework (ESMF); see, for instance, Dickenson et al. (2002).

13. In that sense, one can accurately describe them as parallel rather than serial models, in the sense discussed in Winsberg (2006).

14. "Balance of approximations" is a term introduced by Lambert and Boer (2001) to indicate that climate models sometimes succeed precisely because the errors introduced by two different approximations cancel out.

15. There has been a move, in recent years, to eliminate "legacy code" from climate models. Even though this may have been achieved in some models (this claim is sometimes made about CM2), it is worth noting that there is a large difference between coding a model from scratch and building it from scratch (i.e., devising and sanctioning from scratch all the elements of a model).

16. See especially Biddle and Winsberg (2009) and also chapter 6 of Winsberg (2010).

17. One might complain that if the decisions do not reflect the explicit psychological motives or interests of the scientist, then they do not have a *systematic* effect on the content of science, and are hence no different than the uncontroversial examples of social values I mentioned in the introduction (such as attaching greater value to AIDS research than to algebraic quantum field theory). But though the effect of the values in the climate case might not have a *systematic* effect on the content of science, it

is nonetheless an effect *internal* to science in a way that those other examples are not.

REFERENCES

Allen, Myles. 2008. "What Can Be Said about Future Climate? Quantifying Uncertainty in Multi-Decade Climate Forcasting." ClimatePrediction.net, Oxford University. https://www.climateprediction.net/wp-content/publications/allen_Harvard2008.pdf.

Biddle, Justin, and Eric Winsberg. 2009. "Value Judgments and the Estimation of Uncertainty in Climate Modeling." In *New Waves in the Philosophy of Science,* edited by P. D. Magnus and J. Busch, 172–97. New York: Palgrave MacMillan.

Churchman, C. West. 1948. *Theory of Experimental Inference.* New York: Macmillan.

Churchman, C. West. 1956. "Science and Decision Making." *Philosophy of Science* 22: 247–49.

Dickenson, Robert E., Stephen E. Zebiak, Jeffrey L. Anderson, Maurice L. Blackmon, Cecelia De Luca, Timothy F. Hogan, Mark Iredell, Ming Ji, Ricky B. Rood, Max J. Suarez, and Karl E. Taylor. 2002. "How Can We Advance Our Weather and Climate Models as a Community?" *Bulletin of the American Meteorological Society* 83: 431–34.

Douglas, Heather. 2000. "Inductive Risk and Values in Science." *Philosophy of Science* 67: 559–79.

Dunne, J. 2006. "Towards Earth System Modelling: Bringing GFDL to Life." Presented at the 18th Annual Australian Community Climate and Earth System Simulator (ACCESS) BMRC Modeling Workshop, November 28–December 1, 2006.

Elliot, Kevin C. 2011a. "Direct and Indirect Roles for Values in Science." *Philosophy of Science* 78: 303–24.

Elliot, Kevin C. 2011b. *Is a Little Pollution Good for You? Incorporating Societal Values in Environmental Research.* New York: Oxford University Press.

Frank, Philipp G. 1954. "The Variety of Reasons for the Acceptance of Scientific Theories." In *The Validation of Scientific Theories,* edited by P. G. Frank, 3–17. Boston: Beacon Press.

Gleckler, P. J., K. E. Taylor, and C. Doutriaux. 2008. "Performance Metrics for Climate Models." *Journal of Geophysical Research* 113 (D6): D06104. doi:10.1029/2007JD008972.

Goldstein, M., and J. C. Rougier. 2006. "Bayes Linear Calibrated Prediction for Complex Systems." *Journal of the American Statistical Association* 101: 1132–43.

Howard, Don A. 2006. "Lost Wanderers in the Forest of Knowledge: Some Thoughts on the Discovery-Justification Distinction." In *Revisiting Discovery and Justification: Historical and Philosophical Perspectives on the Context Distinction,* edited by J. Schickore and F. Steinle, 3–22. New York: Springer.

Jeffrey, Richard C. 1956. "Valuation and Acceptance of Scientific Hypotheses." *Philosophy of Science* 22: 237–46.

Kourany, Janet. 2003a. "A Philosophy of Science for the Twenty-First Century." *Philosophy of Science* 70: 1–14.

Kourany, Janet. 2003b. "Reply to Giere." *Philosophy of Science* 70: 22–26.

Küppers, Günter, and Johannes Lenhard. 2006. "Simulation and a Revolution in Modelling Style: From Hierarchical to Network-like Integration." In *Simulation: Pragmatic Construction of Reality,* edited by J. Lenhard, G. Küppers, and T. Shinn, 89–106. Dordrecht, the Netherlands: Springer.

Lambert, Steven J., and G. J. Boer. 2001. "CMIP1 Evaluation and Intercomparison of Coupled Climate Models." *Climate Dynamics* 17: 83–106.

Lenhard, Johannes, and Eric Winsberg. 2010. "Holism, Entrenchment, and the Future of Climate Model Pluralism." *Studies in History and Philosophy of Modern Physics* 41: 253–62.

Longino, Helen. 1990. *Science as Social Knowledge: Values and Objectivity in Scientific Inquiry.* Princeton, N.J.: Princeton University Press.

Longino, Helen. 1996. "Cognitive and Non-Cognitive Values in Science: Rethinking the Dichotomy." In *Feminism, Science, and the Philosophy of Science,* edited by L. H. Nelson and J. Nelson, 39–58. Dordrecht, the Netherlands: Kluwer Academic.

Longino, Helen. 2002. *The Fate of Knowledge.* Princeton, N.J.: Princeton University Press.

Masson, D., and R. Knutti. 2011. "Climate Model Genealogy." *Geophysical Research Letters* 38: L08703. doi:10.1029/2011GL046864.

McMullin, Ernan. 1983. "Values in Science." In *PSA 1982: Proceedings of the Biennial Meeting of the Philosophy of Science,* 1982. Vol. 2: *Symposia and Invited Papers,* 3–28. Chicago: University of Chicago Press.

Neurath, Otto. (1913) 1983. "The Lost Wanderers of Descartes and the Auxiliary Motive (On the Psychology of Decision)." In *Philosophical Papers 1913–1946,* edited and translated by R. S. Cohen and M. Neurath, 1–12. Dordrecht, the Netherlands: D. Reidel. First published as "Die Verirrten des Cartesius und das Auxiliarmotiv. Zur Psychologie des Entschlusses," in *Jahrbuch der Philosophischen Gesellschaft an der Universität Wien* (Leipzig: Johann Ambrosius Barth).

Palmer, T. N. 2001. "A Nonlinear Dynamical Perspective on Model Error: A Proposal for Non-local Stochastic–Dynamic Parameterization in Weather and Climate Prediction Models." *Quarterly Journal of the Royal Meteorological Society* 127: 279–304.

Rudner, Richard. 1953. "The Scientist *Qua* Scientist Makes Value Judgments." *Philosophy of Science* 20: 1–6.

Shackley, Simon, J. Risbey, P. Stone, and Brian Wynne. 1999. "Adjusting to Policy Expectations in Climate Change Science: An Interdisciplinary Study of Flux Adjustments in Coupled Atmosphere Ocean General Circulation Models." *Climatic Change* 43: 413–54.

Shewhart, Walter A. 1939. *Statistical Method from the Viewpoint of Quality Control.* New York: Dover.

Smith, Leonard A. 2002. "What Might We Learn from Climate Forecasts?" *Proceedings of the National Academy of Sciences of the United States of America* 4: 2487–92.

Solomon, Miriam. 2001. *Social Empiricism.* Cambridge, Mass.: MIT Press.

Tebaldi, Claudia, and Reto Knutti. 2007. "The Use of the Multi-model Ensemble in Probabilistic Climate Projections." *Philosophical Transactions of the Royal Society A: Mathematical, Physical and Engineering Sciences* 365 (1857): 2053–75.

Weart, Spencer. 2010. "The Development of General Circulation Models of Climate." *Studies in History and Philosophy of Modern Physics* 41: 208–217.

Wilholt, Torsten. 2009. "Bias and Values in Scientific Research." *Studies in History and Philosophy of Science* 40: 92–101.

Winsberg, Eric. 2006. "Handshaking Your Way to the Top: Simulation at the Nanoscale." *Philosophy of Science* 73: 582–594.

Winsberg, Eric. 2010. *Science in the Age of Computer Simulation.* Chicago: University of Chicago Press.

9

VALIDATING IDEALIZED MODELS

MICHAEL WEISBERG

Cᴌᴀssɪᴄᴀʟ ᴄᴏɴғɪʀᴍᴀᴛɪᴏɴ ᴛʜᴇᴏʀʏ ᴇxᴘʟᴀɪɴs how empirical evidence confirms the truth of hypotheses and theories. But much of contemporary theorizing involves developing models that are idealized—not intended to be truthful representations of their targets. Given that theorists in such cases are not even aiming at developing truthful representations, standard confirmation theory is not a useful guide for evaluating idealized models. Can some kind of confirmation theory be developed for such models?

This discussion takes some initial steps toward answering this question and toward developing a theory of model *validation*. My central claim is that model validation can be understood as confirming hypotheses about the nature of the model/world relation. These hypotheses have truth values and hence can be subject to existing accounts of confirmation. However, the internal structure of these hypotheses depends on contextual factors and scientists' research goals. Thus, my account deviates in substantial ways from extant theories of confirmation. It requires developing a substantial analysis of these contextual factors in order to fully account for model validation. I will also argue that validation is not always the most important kind of model evaluation. In many cases, we want to know if the predictions and explanations (hereafter *results*) of models are reliable. Before turning to these issues directly, I will begin my discussion with two illustrative examples from contemporary ecology.

MODELING EUROPEAN BEECH FORESTS

Prior to extensive human settlement, a forest of beech trees covered much of central Europe. This was a *climax forest,* the result of a steady state of undisturbed forest succession processes. Although most of the forest was composed of thick canopy, where the beech forest was thin trees of various species, sizes, and ages could grow. What caused this patchy pattern?

This is not a question that admits of an easy empirical answer by observation or experiment. The time and spatial scales involved make standard empirical methods difficult to apply, especially if one is hoping to actually reproduce an entire succession cycle. A more practical approach to understanding what caused patchiness is to *model,* to construct simplified and idealized representations of real and imagined forests.

A number of models have been proposed for the climax beech forests of central Europe, and in the rest of this section I will discuss two of them. Both models are *cellular automata.* This means that the model consists of an array of cells, each of which is in a particular developmental state. The model proceeds in time steps whereby each cell on the grid updates its state according to the model's transition rules. In this first model, cells represent small patches of forest, which can be composed of beech, pioneer species such as birch, or a mixture of species.

The first model (hereafter the Wissel model) is a simple cellular automaton based on the following empirically observed pattern of succession. When an old beech tree has fallen, an opening appears, which may be colonized by a first pioneer species such as birch. Then a second phase with mixed forest may arise. Finally, beech appears again, and on its death the cycle starts anew (Wissel 1992, 32). These stages are represented in the model's states and transition rules. A cell can be in any one of these four states (open, pioneer, mixed, mature beech), and each cell will cycle from one state to the next over a defined time course.

If this were the entire content of the Wissel model, it would not be scientifically interesting. Any spatial patterns that it exhibited would be solely a function of the initial distribution of states. If all the cells began in the same state, then the cells would synchronously fluctuate between the four states. If the initial states were distributed randomly, then the forest distribution would appear to fluctuate randomly. The only way to generate a patchy distribution would be to start off from a patchy distribution of states.

To make the model scientifically interesting, Wissel introduced an additional transition rule: the death of one tree can affect its neighbors, especially those to the north.

> If an old beech tree falls the neighbouring beech tree to the north is suddenly exposed to solar radiation. A beech tree cannot tolerate strong solar radiation on its trunk during the vegetation period and dies in the following years. Another reason for the death of the beech tree may be a change of microclimate at ground level which is suddenly exposed to solar radiation. (1992, 32)

This empirical fact is represented in the model in two ways. First, parameterized rules give the probability of a beech tree's death as a function of the location of its dead neighbor and its current age. Second, the death of southerly beeches increases the probability of more rapid succession, especially birch growth.

As we can see in Figure 9.1, when the model is initialized with a random distribution of forest stages, the array quickly becomes patchy, with mixtures of tree age and tree species at different grid locations. The author of the study attributes this to local interactions, describing the system as *self-organizing*. "This spatial pattern is the consequence of a self-organization in the system. It is caused by the interaction of neighbouring squares, i.e. by the solar radiation effect" (42). In other words, patchiness in real forests is an emergent property controlled by the succession cycle and the solar radiation effect.

The second model I am going to describe is considerably more complex, although still extremely simple relative to real-world forests. The BEFORE (BEech FOREst) model builds on some ideas of the Wissel model but adds an important new dimension: height (Rademacher et al. 2004). Succession patterns do not just involve changes in species but changes in the vertical cross section of a forest. Early communities contain short plants and trees. By the time mature beech forest is established, the canopy is very high, and the understory is mostly clear. The tallest trees receive all of the sunlight because of this canopy. At the same time, the height of these mature beech trees makes them susceptible to being damaged or killed by powerful wind storms.

To represent height, the BEFORE model stacks four cells vertically on top of each cell of the two-dimensional grid. These vertical cells correspond to the ground (seedling) layer, the juvenile tree layer, the lower canopy, and

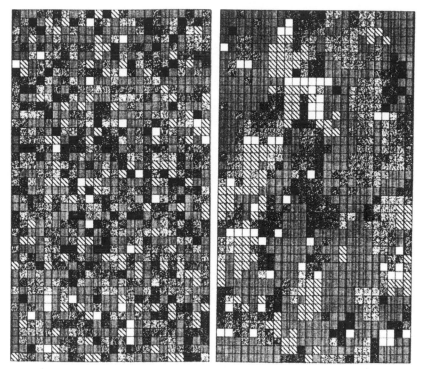

Figure 9.1. The distribution of succession phases in the Wissel model. On the left is a random distribution of phases. On the right is a "patchy" distribution after 170 time steps of the model. (From Wissel 1992, 33 and 36; reprinted with permission of Elsevier.)

the upper canopy. Each vertical cell can contain trees of the appropriate height, with trade-offs restricting the total number of trees that can grow at a particular cell on the two-dimensional lattice.

Another difference between the two models is that in BEFORE the trees in the upper canopy are represented as having crowns. This is a significant addition because it allows the model to represent the interaction between crown widths, the sizes of other trees, and the amount of sunlight reaching the forest floor. For example, a single tree that takes up 50 percent of a cell's horizontal area will prevent more trees from growing in a given cell than one that takes up 12.5 percent. Spaces in the upper canopy allow trees in the lower canopy to rapidly grow and fill this space. Spaces in both canopies allow light to reach the understory and promote growth. Further, the model allows an occasional heavy wind to topple the tallest trees and hence to open spaces in the canopy. A simplified representation of the BEFORE model is shown in Figure 9.2.

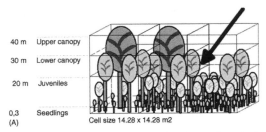

Figure 9.2. A simplified representation of the BEFORE model showing hypothetical populations of three-dimensional grid elements. (From Rademacher et al., 2004, 352. Courtesy of V. Grimm.)

Like the Wissel model, simulations using the BEFORE model yield mosaic patterns of trees. The greater sophistication of the BEFORE model, however, allows for more elaborate patterns to emerge. For example, one interesting result concerns the kinds of disturbances that disrupt the mosaic patterns. When an extremely strong storm is unleashed in the model, destroying large parts of the upper canopy, the forest quickly homogenizes in the damaged locations. "It takes about seven time steps (ca. 100 years) until the effect, or 'echo,' of the extreme storm event is 'forgotten,' i.e. no longer detectable in the forest structure and dynamics" (Rademacher et al. 2004, 360).

Another interesting result of the BEFORE model concerns the synchrony of the succession dynamics. With this model, we can ask whether some particular forest location is in more or less the same developmental stage as another. We find that local dynamics are synchronized, but global dynamics are not. "Development is synchronised at the local scale (corresponding to few grid cells), whereas at the regional scale (corresponding to the entire forest) it is desynchronised" (360). Interestingly, storms are the overarching causal factor leading to local synchrony. "The synchronising effect of storms at the local scale is due to the damage of the windfalls. This damage 'spreads' into the neighbourhood so that the state and, in turn, the dynamics of neighbouring cells become partly synchronised" (361). This means that it is an exogenous factor, storms, not an endogenous process that leads to the observed synchronization. Such a mechanism was not predicted by the Wissel model, nor could it have been because the Wissel model does not represent factors exogenous to the forest/sun system.

Rademacher and colleagues are also quite explicit about why, despite the high degree of idealization of their BEFORE model, they take it to be validated:

When designing BEFORE, a model structure was chosen which in principle allows two patterns to emerge but which were not hard-wired into the model rules: the mosaic pattern of patches of different developmental stages, and the local cycles of the developmental stages. The question was whether the processes built into the model are sufficient to reproduce these two patterns. The results show that indeed they are: in every time step, the same type of mosaic pattern as observed in natural forests . . . is produced. (2004, 363)

They go on to explain how they subjected this model to *sensitivity analysis,* or what philosophers of science have often called *robustness analysis* (Levins 1966; Wimsatt 1981, 2007; Weisberg 2006b; Woodward 2006).

The BEFORE and Wissel models are good examples of the type of model that plays a role in contemporary theoretical research. Relative to their intended target systems, these models contain many simplifications and idealizations. Yet despite these distortions, ecologists still regard these models as explanatory. They see the models as successful contributions to the ecological literature, ones that deepen our understanding of forest succession.

The ways in which these models were assessed deviates in many ways from philosophical accounts of theory confirmation. The most important deviation concerns truth. As standardly developed, accounts of confirmation show how to assess the truth of a theory given the available evidence. But neither of these models is even a remotely true description. Before we look at any empirical evidence, we know they are not accurate representations of a real forest's ecology.

Moreover, despite it being a better model, it is misleading to say that the BEFORE model is a more accurate representation of a forest than the Wissel model. Not only are both of these models very far from the truth, they are not even intended to represent all the same things. For example, it would be wrong to say that the Wissel model incorrectly represents the vertical structure of forests; it does not even try to represent it.

Nevertheless, theorists judged the BEFORE model, and to a lesser extent the Wissel model, to be at least partially validated empirically (Turner 1993). Moreover, the BEFORE model was judged to be an improvement of the Wissel model. So if we reject the idea that model validation has the aim of determining the truth of models, what does it aim to tell us? I will endeavor to answer this question in the next two sections.

VALIDATION AS SEEN BY PRACTICING MODELERS

Although the philosophical literature about validating idealized models is not well developed, a number of practicing modelers have discussed the issue. In this section, I will discuss a couple of key passages from this literature, focusing on ideas that point the way toward an account of model validation.

In understanding many aspects about the nature of modeling, it is useful to look at the discussions of Vito Volterra. Volterra was a mathematical physicist who began addressing biological problems in response to specific, practical questions about fishery dynamics after World War I. His work on Italian fisheries proved highly influential and helped to inaugurate modern mathematical biology (Kingsland 1995; Weisberg 2007). In describing his approach to mathematical biology, Volterra explained how he began working on a new theoretical problem:

> As in any other analogous problem, it is convenient, in order to apply calculus, to start by taking into account hypotheses which, although deviating from reality, give an approximate image of it. Although, at least in the beginning, the representation is very rough . . . it is possible to verify, quantitatively or, possibly, qualitatively, whether the results found match the actual statistics, and it is therefore possible to check the correctness of the initial hypothesis, at the same time paving the way for further results. Hence, it is useful, to ease the application of calculus, to schematize the phenomenon by isolating those actions that we intend to examine, supposing that they take place alone, and by neglecting other actions. (Volterra 1926, 5; translation G. Sillari)

This comment suggests that the competent modeler does not attempt to create and validate a single, best model in one shot. Instead, Volterra suggested that modelers should begin by making educated guesses about the essential quantities and interactions required in a model. The modeler then checks qualitatively, or quantitatively if possible, whether the model is sufficient for her purposes. If not, which will be most of the time, she refines the model, adding additional factors. This process continues as the modeler learns more about the target system, as technology for modeling improves, and as our needs change.

This passage is a good summary of the modeling cycle (Grimm and Railsback 2005), the steps involved in creating, analyzing, and testing a model. It is important to emphasize, as Volterra did, that these steps happen over time and are iterative. But Volterra leaves many questions unanswered, including, What guides the choice of model? How are the qualitative and quantitative features of models compared to their targets?

Most modelers who have written about modeling see the choice of model as having a substantial pragmatic element. They do not necessarily dispute that the relationship between a model and its target can be determined objectively, just that the choice of a particular model depends on the interests of the scientific community. For example, chemists Roald Hoffmann, Vladimir Minkin, and Barry Carpenter write,

> Real chemical systems, be they the body, the atmosphere, or a reaction flask, are complicated. There will be alternative models for these, of varying complexity. We suggest that if understanding is sought, simpler models, not necessarily the best in predicting all observables in detail, will have value. Such models may highlight the important causes and channels. If predictability is sought at all cost—and realities of marketplace and judgments of the future of humanity may demand this—then simplicity may be irrelevant. (1997, 17)

The key of this analysis is that simple or highly idealized models may be especially useful for explanatory purposes, even when they are not ideal for making predictions. But why might this be? Would not a better model be one that was explanatory and predictive?

Richard Levins explored one possible explanation for the decoupling of explanatory and predictive models in his landmark study "The Strategy of Model Building in Population Biology" (1966). He argued that scientists must make decisions about which strategy of model building to employ because there are *trade-offs* among three desiderata of modeling: precision, generality, and realism (Levins 1966). Implicit in his argument was the idea that highly general models are more explanatory because they support counterfactuals. But when a scientist needs to make a prediction, she should aim for greater degrees of what Levins called "realism."

While the specific trade-off Levins discusses is controversial (Orzack and Sober 1993; Weisberg 2006a; Matthewson and Weisberg 2009), the consensus in the modeling and philosophical literature is that some trade-offs are real and that that they may constrain modeling practice in ways that require the generation of targeted models. A related consideration discussed by Levins, and elaborated on by William Wimsatt, is that many targets are simply too complex to be captured in a single model. Different models will be useful for different purposes, and it is only collections of models that can guide scientists to making true statements about their targets (Wimsatt 1987, 2007; see Isaac 2013 for a more pragmatic reconstruction of the Levins/ Wimsatt argument).

The second major theme in Volterra's discussion of model building has to do with the validation of the model itself. Specifically, he says that verification is connected to the ways in which the model's results qualitatively and quantitatively correspond to properties of a real system. Surprisingly little has been written about this issue. The classic literature about models and modeling emphasizes the discovery of structural mappings such as *isomorphism* and *homomorphism* (e.g., Suppes 1962; van Fraassen 1980; Lloyd 1994). These take the form of exact structural matches, not approximate qualitative or quantitative matches. There are, of course, formal statistical tests that show the goodness-of-fit between a mathematical model and a data set. But again, these are not especially helpful for assessing a qualitative fit between an idealized model and a complex real phenomenon. Helpfully for us, practicing modelers have addressed some aspects of this issue.

One of the clearest discussions of qualitative and quantitative fit can be found in a study by Volker Grimm and colleagues (2005). Like Volterra, Grimm and coworkers put a great deal of emphasis on initial qualitative testing. They describe a procedure of testing models against a set of prespecified output patterns drawn from empirical sources. "Models that fail to reproduce the characteristic patterns are rejected, and additional patterns with more falsifying power can be used to contrast successful alternatives" (Grimm et al. 2005, 988). One continually makes small changes to a model until eventually the pattern can be reproduced. At that point, one can choose another pattern and then repeat the process.

One of the important things that Grimm et al. emphasize is that the patterns to which models should be compared are often qualitative. They give an example from social science:

In a model exploring what determines the access of nomadic herdsmen to pasture lands owned by village farmers in north Cameroon, herdsmen negotiate with farmers for access to pastures . . . Two theories of the herdsmen's reasoning were contrasted: (i) "cost priority," in which herdsmen only consider one dimension of their relationship to farmers—costs; and (ii) "friend priority," in which herdsmen remember the number of agreements and

refusals they received in previous negotiations. Real herdsmen sustain a social network across many villages through repeated interactions, a pattern
reproduced only by the "friend priority" theory. (2005, 989)

They go on to discuss how, with further refinements, such qualitative patterns
can be turned into quantitative ones. Ultimately, statistics can be used to
verify the match between a quantitative pattern and modeling results.
Grimm et al. describe this as a cyclical process that simultaneously leads to
discovering more about the world and getting a better model of the world.

This brief look at the practitioners' literature on model verification has
highlighted several of the themes that will guide my subsequent discussion.
The first theme, which we see in Volterra as well as in Grimm et al., is that
model validation is an iterative process. A second theme is that both qualitative and quantitative agreements are important, so it will not do to regard
the problem of model validation as one that can be settled by simple statistical tests. Finally, and perhaps most importantly, all these authors see a connection between the pragmatics of modeling and epistemic questions about
validation. These authors think that there are different kinds of models for
different purposes, so there will rarely be a single, all-purpose model.

THEORETICAL HYPOTHESES

Having looked at some of the major issues involved in model verification, as
enumerated by working modelers, I want to turn back to the philosophical
literature about confirmation. It is evident that the traditional conception of
confirmation, the attempt to ascertain the truth (or empirical adequacy) of
a hypothesis, cannot be directly applied to the validation of models. This is
the case for several reasons. Models are often highly idealized: they may not
even be intended to quantitatively take into account the details of a real-
world system, their evaluation is connected to the pragmatics of research,
and multiple models of the same target system may generated without
apparent conflict. All these factors prevent model validation from simply
being a matter of estimating a model's truth or accuracy.

Rather than connecting validation to truth or accuracy, a number of philosophers have suggested that the central model/world relation is *similarity*
(Giere 1988). Scientifically useful models are not meant to be truthful representations but rather are meant to be similar to their targets in certain respects and degrees. If models are similar to their targets, then an account of

validation needs to show how evidence bears on the similarity relation, not the truth relation. Specifically, it needs to account for how evidence shows that a model is similar to its target.

One way to develop such an account is to formulate hypotheses about a model's similarity to its target. This approach is suggested, if not endorsed, by Giere (1988), who says that scientists are interested in the truth not of models but of *theoretical hypotheses,* which describe the similarity relationship between models and the world. What form should such theoretical hypotheses take? Following Giere, we can formulate them as follows:

Model M is similar to target T in respects X, Y, and Z.   (Eq. 9.1)

Such a hypothesis has an informative truth value. Hence, ideas from classical confirmation theory can be used to evaluate this hypothesis.

Reduction of validation to confirmation via theoretical hypotheses seems fruitful. It potentially lets us reduce a very difficult problem to one that we already know something about. However, there are three major issues that will need to be addressed in order to develop a working account of model validation. First, we need to have an account of the similarity relation. Second, given that every model is similar to every target in some respect or other, we need a principled way of specifying which respects are the relevant ones. Finally, we need to take into account the pragmatic dimension of modeling. Specifically, we need to know what purpose the model was intended for and how this constrains the choice of similarity relation.

In the remainder of this section, I will explain how these issues can be addressed. I will begin by giving a more formal account of theoretical hypotheses. Then in subsequent sections I will discuss how this account helps us specify the relevant degrees of similarity and take into account the pragmatic dimensions of modeling.

Weighted Feature Matching

In *Simulation and Similarity* (2013), I argue that the best account of the model/world similarity relation is *weighted feature matching,* which is based on ideas from Amos Tversky (1977). This account begins with a set of features of the model and the target, which I call the *feature set* or Δ. This set contains static and dynamic properties, which I call *attributes,* and causal features, which I call *mechanisms.* Sets of the model's attributes from Δ are labeled M_a and of the target's attributes T_a. Similarly, the model's mechanistic features from Δ are labeled M_m and of the target's mechanistic features are labeled T_m.

What features can be included in Δ? Following scientific practice, I think we should be very liberal about what is allowed in Δ. This will include qualitative, interpreted mathematical features such as "oscillation," "oscillation with amplitude A," "the population is getting bigger and smaller," and strictly mathematical terms such as "is a Lyapunov function." They can also include physically interpreted terms such as "equilibrium," "average abundance," or "maximum tree height."

The overall picture is that the similarity of a model to its target is a function of the features they share, penalized by the features they do not share. To normalize the similarity relation, this is expressed as the ratio of shared features ($M_a \cap T_a$ and $M_m \cap T_m$) to nonshared features ($M_a - T_a$ and so forth). To generate numbers from sets such as ($M_a \cap T_a$), we need to introduce a weighting function f, which is defined over a powerset of Δ. Once this is completed, we can write the similarity expression as follows:

$$S(m,t) = \frac{\theta f(M_a \cap T_a) + \rho f(M_m \cap T_m)}{\theta f(M_a \cap T_a) + \rho f(M_m \cap T_m) + \alpha f(M_a - T_a) + \beta f(M_m - T_m) + \gamma f(T_a - M_a) + \delta f(T_m - M_m)}$$

(Eq. 9.2)

If the model and target share all features from Δ, then this expression is equal to 1. Conversely, if they share no features, then this expression is equal to 0. But the interesting and realistic cases are when they share some features; then the value of S is between 0 and 1. In such cases, this expression is useful for two purposes. For a given target and scientific purpose, the equation lets us evaluate the relative similarity of a number of models by scoring them. Moreover, when multiple plausible models have been proposed, this expression helps us isolate exactly where they differ.

For example, the Wissel model shares the attribute of patchiness with its targets. This attribute can be specified very abstractly (e.g., mere patchiness) or can be refined either qualitatively (e.g., moderate patchiness) or quantitatively (e.g., clusters with average extent 3.4 units). The model also shares the succession pattern itself, another attribute, with its target. On the other hand, because it is a two-dimensional model, attributes along the vertical dimension of the forest are not represented at all in the model.

Giere-like theoretical hypotheses can be generated with the weighted feature matching expression. They have the following form:

Model m is similar to target t for scientific purpose p to degree $S(m, t)$.

(Eq. 9.3)

Validation of idealized models is then aimed at gathering evidence that sentences in the form of this second expression (equation 9.3) are true. But more importantly, when such a sentence is disconfirmed, theorists can use the information in the first expression of this chapter (equation 9.1), at least in principle, to help them discover where the deviation is taking place.

If we accept that the second expression (equation 9.3) gives the correct form of theoretical hypotheses, we need to work out how we can fill in the p term. Specifically, we need to understand the relationship between the scientific purpose p and the degree of similarity.

Pragmatics and Weighting Terms

One advantage of articulating theoretical hypotheses using the weighted feature matching formalism is that we can say a lot more about what is meant by scientific purpose p. Specifically, the weighting function f and the relative weight (or inclusion at all) of the terms in the first expression in the chapter are contextual factors reflecting the scientific purpose.

For example, if a theorist is constructing a model to be deployed in a how-possibly explanation, she is interested in constructing a model that reproduces the target's static and dynamic properties by some mechanism or other. This means that a good model, in this context, is one that shares many and does not lack too many of the target's static and dynamic properties. Analysis of the model's properties and evidence about the target's properties are brought together to make this determination. This corresponds to $|M_a \cap T_a|$ having a high value and $|M_a - T_a|$ having a low value. We can formalize this idea by saying that in the construction of how-possibly explanations theorists aim for the following expression to equal 1:

$$\frac{f(M_a \cap T_a)}{f(M_a \cap T_a) + f(M_a - T_a)}$$

(Eq. 9.4)

As we often fall short of such a goal, this expression (equation 9.4) allows us to compare models to one another in the following ways. First, we can see which features, among the ones that matter, are omitted by different models. Second, assuming a common feature set and weighting function, the models' relative deviation from the target can be assessed.

This also suggests how theoretical hypotheses for how-possibly modeling are formulated. One starts from the second expression in this chapter, and substitutes the expression in equation 9.4. The resulting theoretical hypothesis can be formulated as follows:

How-possibly model m is similar to target t to degree:

$$\frac{f(M_a \cap T_a)}{f(M_a \cap T_a) + f(M_a - T_a)} \qquad \text{(Eq. 9.5)}$$

Confirmation of a how-possibly model thus amounts to demonstrating that the model shares many of the target's attributes. It does not, however, involve showing that the target shares the model's mechanisms.

Other scientific purposes will involve the construction of minimal models rather than how-possibly models. In this type of modeling, theorists want to find one or a small number of mechanisms thought to be the first-order causal factors giving rise to the phenomenon of interest, but nothing else. When building such a model, the theorist wants to ensure that all the mechanisms in her model are in the target (high $|M_m \cap T_m|$), that her model reproduces the first-order phenomena of interest (high $|M_a \cap T_a|$), and that there are not any extraneous attributes or mechanisms in the model (low $|M_m - T_m|$ and $|M_a - T_a|$). On the other hand, it is perfectly fine for the target to have mechanisms and attributes not in the model. This corresponds to the following expression having a value near one:

$$\frac{f(M_m \cap T_m)}{f(M_m \cap T_m) + (M_m - T_m) + f(M_m - T_m)} \qquad \text{(Eq. 9.6)}$$

It also corresponds to the truth of the statement $\exists \psi \in \Delta, \psi \in (M_m \cap T_m)$, and ψ is a first-order causal factor of T. So a theoretical hypothesis for a minimal model would be written as follows:

A minimal model m is similar to target t to degree:

$$\frac{f(M_m \cap T_m)}{f(M_m \cap T_m) + (M_m - T_m) + f(M_m - T_m)} \qquad \text{(Eq. 9.7)}$$

and $\exists \psi \in \Delta, \psi \in (M_m \cap T_m)$, and ψ is a first-order causal factor of T.

Other expressions could be generated for different scientific purposes along the same lines. For example, hyperaccurate modeling would use the first expression in this chapter. Mechanistic modeling, where the goal is to understand the behavior of some mechanism in the world, would involve a high degree of match between the model's representation of mechanisms and the target's mechanisms.

Now that we have some idea of how theoretical hypotheses can be constructed to take into account scientific purposes, there still remains the difficult question of feature weighting. Depending on scientific goals and other contextual factors, elements of Δ will be weighted differently. How is that determined?

Relevance and Weighting Function

Mathematically, weights are assigned to individual features by the weighting function f. Formally, this function needs to be defined for every element in $\wp(\Delta a) \cup \wp(\Delta m)$. However, it is very unlikely that scientists even implicitly construct functions defined over this set, or even could produce such a function if called on to do so. For nontrivial Δs, there are simply too many elements in this powerset. Moreover, it is unlikely that the relative importance of features would be thought of in terms of the elements of such a powerset. Rather, scientists typically think about the relative weights of the individual elements of Δ. This means that we should restrict the weighting function by requiring that:

$$f\{A\} + f\{B\} = f\{A, B\} \qquad \text{(Eq. 9.8)}$$

In other words, the weight of a set is equivalent to the weight of each element of the set.

Even with this restriction, the weighting function requires a weight for each element in Δ, something scientists are unlikely to have access to. More realistically, scientists will think about weighting in the following way. Some subset of the features in Δ is especially important, so let us call this subset the set of *special features*. These are the features that will be weighted more heavily than the rest, while the nonspecial features will be equally weighted. So as a default, the weighting function will return the cardinality of sets like $(M_m \cap T_m)$. The subset of special features will receive higher weight according to their degree of importance.

This adds further constraints on the weighting function, but we still are left with two questions: How do scientists determine which elements of Δ are the special features? What weights should be put on them? In some cases, a fully developed background theory will tell us which terms require the greatest weights. For example, complex atmospheric models are ultimately based on fluid mechanics, and this science can tell us the relative importance of different features of the target for determining its behavior.

Unfortunately, many cases of modeling in biology and the social sciences will not work this straightforwardly. In these cases, background theory will not be rich enough to make these determinations, which means that the basis for choosing and weighting special features is less clear. In such cases, choosing an appropriate weighting function is in part an empirical question. The appropriateness of any particular function has to be determined using means-ends reasoning.

To return to an example, in constructing the BEFORE model, Rademacher and colleagues had the goal of "constructing an idealised beech forest revealing typical general processes and structures." This led them to put little weight on, and indeed not include in the model, features related to "the growth, demography or physiology of individual beech trees or specific forest stands [of trees]" (Rademacher et al. 2004, 351). Because of their focus, they put special emphasis on the model being able to reproduce the mature forest attribute of a more or less closed canopy with little understory. In addition, they aimed for their model to reproduce the heterogeneous ages or developmental stages of the various patches.

For this particular model, the theorists simply did not know which mechanisms were important. Instead, Rademacher and coworkers used the their model to generate evidence about which mechanisms cause patchiness. They intended to give high weight to the first-order causal mechanisms that actually gave rise to these attributes. But only empirical and model-based research could reveal what these mechanisms were.

This example nicely illustrates a major theme of this book: modeling practice involves an interaction between the development of the model and the collection of empirical data. As these practices are carried out in time, scientists can discover that they had mistakenly overweighted or underweighted particular features of their models.

EVIDENCE AND THEORETICAL HYPOTHESES

Once a modeler has an idea about the theoretical hypothesis she is interested in confirming, she can gather the data needed to verify it. As I have already said, this may sometimes involve modifying the theoretical hypothesis in light of what is already known. But say that we are in the position of having a satisfactory model and a satisfactory theoretical hypothesis. In that case, how are data brought to bear on this hypothesis?

Following the form of the theoretical hypothesis, the theorist has two basic tasks. She needs to determine which attributes are shared between target and model, and which ones are not shared. This task can be simplified and focused in a couple of ways. First, there will often be aspects of the model that are known from the start to not be shared by the target. These can be physical idealizations (e.g., representing a finite population as infinite, or representing space as toroidal) or artifacts of the mathematics (e.g., noninteger population sizes).

Sometimes these features, especially the mathematical artifacts, will be excluded from the initial feature set. But this is not always possible or even desirable. Often a theorist will specifically introduce these misrepresentations to compensate for some other misrepresentation of the target, in which case they would be included in the feature set. In any case, intentional distortions can automatically be included among those features that are not shared between model and target.

Another group of features that can be counted in the $(M - T)$ set are all those features that are absent from the model by virtue of its being more abstract than the target. As Elliott-Graves (2013) points out, much of scientific abstraction takes place in the choice of the target. Both the Wissel model and the BEFORE model have forest targets that do not include the color of the leaves in the mature tree canopy, for example. So the target itself is abstract relative to real European forests. But models can contain further abstractions. Although they share a target, the Wissel model is more abstract than the BEFORE model because it does not represent trees as having heights; it is two-dimensional. Any features that are not represented in the model because they are excluded by abstraction are counted in the $(M - T)$ set.

Finally, we come to the heart of the matter, the features that are determined to be in $(M - T)$ or $(M \cap T)$ by empirical investigation. How do theorists empirically determine when features are shared between models and targets?

Part of the answer to this question is simple: you have to "do science." Observations and experiments are required to determine facts about the target system. These facts about the target can then be compared with the model. Because there are no general recipes for empirical investigation, I cannot give an informative account about it in this essay. However, I can say something about how the target system needs to be conceptualized such that it can be compared to the model.

Following Elliott-Graves (2013), I see target systems as *parts* of real-world phenomena. Theorists make a choice about which mechanisms and attributes they wish to capture with their model, and then they try to construct models to take into account these mechanisms and attributes. In order for this to work, like has to be compared with like. Physical models, like their targets, are composed of physical attributes and mechanisms. So physical models can be directly compared to real-world targets. But things are much more complicated for mathematical and computational models because these do not possess physical properties and physical attributes. So we need one additional step in order to compare them to targets.

Mathematical and computational models, as well as concrete models in some cases, are compared to mathematical representations of targets, not the targets themselves. Each state of the target is mapped to a mathematical space. In simple, dynamical models, the mapping is such that the major determinable properties (e.g., canopy height, solar radiation flux, and tree distribution) of the target are mapped to dimensions of a state space, and specific states are mapped to points in this space. Now one interpreted mathematical object (the model) can be compared to another (the mathematical representation of the target), and we avoid problems about comparing mathematical properties to concrete properties. The construction of such a mathematical space is very similar to what Suppes (1962) called a *model of data*.

When one is actually trying to determine the cardinality of $(M \cap T)$, and hence the degree of overlap between model and target, it is typically impractical or impossible to investigate the entirety of T. Theorists sample strategically from this set, often using procedures such as the one Grimm and coworkers (2005) describe as *pattern oriented modeling*. The idea is to look for some characteristic, signature patterns of the target and see if they are reproduced in the model. Or it could go the other way around. One might discover, say, that a forest model always predicted a certain spatial distribution of patchiness; then one would look for this pattern in the target. Another interesting pattern exhibited by the BEFORE model is local synchronization. While the whole forest may be at diverse developmental stages, small patches, corresponding to a few grid cells, will be at the same stage. These patterns and others are the kinds of evidence that are needed to infer some value for $f(M \cap T)$, and hence to confirm a theoretical hypothesis of the form given in equation 9.3.

The development of the model and the empirical investigation of the target are not strictly separable processes. One does not simply find out the contents of the set *M,* then *T,* and then look for overlap. Instead, models often help theorists learn more about what aspects of the target to investigate. Matches between attributes of model and attributes of target may lead theorists to the next empirical phenomenon to investigate. Failures to match may lead either to refinement of the target or, if such failures are deemed acceptable, to the refinement of the theoretical hypothesis.

VALIDATING MODELS AND CONFIRMING THEOREMS

Model validation establishes the extent to which a model is similar to a target system. Although this is important to the modeling enterprise, it is not the only confirmation-theoretic issue that modelers confront. In many instances of modeling, validation is just the first step toward establishing what modelers care about most: the reliability of specific modeling results. Such results are what allow scientists to use models to make predictions about the future, explain the behavior of real targets, and engage in model-based counterfactual reasoning. We might want to know, for example, how much climate change will affect forest productivity. Or perhaps we want to explain the patchiness of forests. Or perhaps we want to know how quickly a forest will recover under different controlled burning protocols.

To make things concrete, let us once again consider the BEFORE model. As I discussed in the first section, this model makes two very important predictions: that forests will take about 100 years to recover from a major storm and that forest development will be locally synchronized. Even if similarity between this model and real European beech forests is established, what licenses the inference from "the model predicts 100 year recoveries and local synchrony" to "100 year recoveries and local synchrony will happen in a real European forest"? Specifically, as the model is not an accurate representation of any real forest, what justifies making predictions and explanations on the basis of it?

Whatever inferences are licensed from the results of the model to the target system have to do, in part, with the extent to which the model is validated. Such inferences seem fully justified when one has a fully accurate model, but let us consider the differences between a fully accurate representation and one that is merely similar to a target in certain respects and degrees. With an accurate representation, one is warranted to believe that mathematically and logically valid analytical operations performed on this representation generate results that transfer directly to the target. This is because, in accurate representations, the causal structure of the target is portrayed accurately using mathematics or some other system with high representational capacity. But this is often not the case with idealized models. We know from the start that idealized models are not accurate representations of their targets, so how does this inference work?

The first step required to license inferences from the results of the model to the properties of a target is model validation. In order to make any inferences from model to target, one needs to learn about the relationship between the model and the target. Moreover, by formulating theoretical hypotheses as I described in the third section, the theorist knows which kinds of features of the model are similar to the target. These play an essential role because they point to the parts of the model that contain the most accurate representations of a target. If these parts of the model are sufficiently similar to their target, and if the mathematics or computations underlying the model have sufficient representational capacity, then operations on these parts of the model are likely to tell us something about the target.

Although validation allows the theorist to isolate the parts of the model that are reliable, and hence the kinds of results that are dependable, uncertainty still remains. Specifically, because one knows ahead of time that the models contain distortions with respect to their targets ("idealizing assumptions"), it may be those idealizations and not the similar parts of the model that are generating a result of interest. Thus, if one wants to establish that a result of interest is a *genuine result*—one that is not driven by idealizing assumptions alone— further analysis is needed. Richard Levins (1966) and William Wimsatt (1987, 2007) called this kind of analysis robustness analysis (see also Weisberg 2006b; Weisberg and Reisman 2008).

Levins originally described robustness analysis as a process whereby a modeler constructs a number of similar but distinct models looking for common predictions. He called the results common to the multiple models *robust theorems*. Levins's procedure can also be centered on a single model, which is then perturbed in certain ways to learn about the robustness of its results. In other words, if analysis of a model gives result *R*, then theorists can begin to systematically make changes to the model, especially to the idealizing assumptions. By looking for the presence or absence of *R* in the new models, they can establish the extent to which *R* depended on peculiarities of the original model.

Three broad types of changes can be made to the model, which I will call *parameter robustness analysis, structural robustness analysis,* and *representational robustness analysis* (Weisberg 2013). Parameter robustness involves varying the settings of parameters to understand the sensitivity of a model result to the parameter setting. This is often called *sensitivity analysis* (Kleijnen 1997) and is almost always a part of published models. The authors of the BEFORE model, for example, conducted parameter robustness analysis for fourteen of the model's sixteen parameters (Rademacher et al. 2004, table 2). Each of these parameters was centered on an empirically observed value but then varied within a plausible range.

The second type of change is called structural robustness analysis, or structural stability analysis (May 2001). Here the causal structures represented by the model are altered to take into account additional mechanisms, to remove mechanisms, or to represent mechanisms differently. In several well-studied cases, adding additional causal factors completely alters a fundamental result of the model. For example, the well-studied undamped oscillations of the Lotka–Volterra model become damped with the introduction of any population upper limit, or carrying capacity imposed by limited resources (Roughgarden 1979).

Finally, a third type of change to the model, called representational robustness analysis, involves changing its basic representational structure. Representational robustness analysis lets us analyze whether the way our models' assumptions are represented makes a difference to the production of a property of interest. A simple way to perform representational robustness analysis is by changing mathematical formalisms, such as changing differential equations to difference equations. In some cases, this can make a substantial difference to the behavior of mathematical models. For example, the Lotka–Volterra model's oscillations cease to have constant amplitude when difference equations are substituted for differential equations (May 1973). In the most extreme forms of representational robustness analysis, the type of model itself is changed. For example, a concrete model might be rendered mathematically, or a mathematical model rendered computationally.

After scientists have established a model's validation and have conducted robustness analysis, where are they left? Are they in a position to be confident of the model's results when applied to a real-world target? I think the answer is a qualified yes. To see this a little more clearly, we can break the inference from model result to inference about the world into three steps.

1. **Discover a model result R.** This licenses us to make a conditional inference of the form "*Ceteris paribus*, if the structures of model M are found in a real-world target T, then R will obtain in that target."

2. **Show the validation of model *M* for target *T*.** This licenses us to make inferences of the form "*Ceteris paribus, R* will obtain in *T*."

3. **Do robustness analysis.** This shows the extent to which the *R* depends on peculiarities of *M*, which is much of what drives the need for the *ceteris paribus* clause. If *R* is highly robust, then a cautious inference of the form "*R* will obtain in *T*" may be licensed.

Thus, by combining validation and robustness analysis, theorists can use their models to conclude things about real-world targets.

I began this chapter by asking a simple question: Can a confirmation theory be developed for idealized models? The answer to this question was by no means certain because when theorists construct and analyze idealized models, they are not even aiming at the truth. Although I only present a sketch of such a theory, I think this question can be answered in the affirmative. But we really need two confirmation-theoretic theories: a validation theory for models and a confirmation theory for the results of models.

Establishing the similarity relationship between models and their targets validates such models. This similarity relation specifies the respects and degrees to which a model is relevant to its targets. The weighting of these respects, which I have called "features," is determined by scientific purposes in ways discussed in the third section. Validation is not the same thing as confirmation, so once we have validated a model, there is still much work to be done before inferences can be drawn from it. Thus, modelers also have to pay special attention to the results of their models, deploying robustness analysis and other tests to determine the extent to which model results depend on particular idealizing assumptions.

Our judgments that models like BEFORE and Wissel are good models of their targets and make reliable predictions does not depend on the truth or overall accuracy of these models. Such properties are conspicuously absent from modeling complex phenomena. But through a process of systematic alignment of model with target, as well as robustness analysis of the model's key results, much can be learned about the world with highly idealized models.

NOTE

Many thanks to Marie Barnett, Shereen Chang, Devin Curry, Alkistis Elliot-Graves, Karen Kovaka, Alistair Issac, Ryan Muldoon, and Carlos

Santana for helpful comments on an earlier draft of this essay. This research was sponsored, in part, by the National Science Foundation (SES-0957189).

REFERENCES

Elliott-Graves, Alkistis. 2013. "What Is a Target System?" Unpublished manuscript.

Giere, Ronald N. 1988. *Explaining Science: A Cognitive Approach.* Chicago: University of Chicago Press.

Grimm, Volker, and Steven F. Railsback. 2005. *Individual-Based Modeling and Ecology.* Princeton, N.J.: Princeton University Press.

Grimm, Volker, Eloy Revilla, Uta Berger, Florian Jeltsch, Wolf M. Mooij, Steven F. Railsback, Hans-Hermann Thulke, Jacob Weiner, Thorsten Wiegand, and Donald L. DeAngelis. 2005. "Pattern-Oriented Modeling of Agent-Based Complex Systems: Lessons from Ecology." *Science* 310 (5750): 987–91.

Hoffmann, Roald, Vladimir I. Minkin, and Barry K. Carpenter. 1997. "Ockham's Razor and Chemistry." *Bulletin de la Société Chimique de France* 133 (1996): 117–30 ; reprinted in *HYLE—International Journal for Philosophy of Chemistry,* 3: 3–28.

Isaac, Alisair M. C. 2013. "Modeling without Representation." *Synthese* 190: 3611–23.

Kingsland, Sharon E. 1995. *Modeling Nature: Episodes in the History of Population Ecology.* Chicago: University of Chicago Press.

Kleijnen, Jack P. C. 1997. "Sensitivity Analysis and Related Analyses: A Review of Some Statistical Techniques." *Journal of Statistical Computation and Simulation* 57 (1–4): 111–42.

Levins, Richard. 1966. "The Strategy of Model Building in Population Biology." *American Scientist* 54: 421–31.

Lloyd, Elizabeth A. 1994. *The Structure and Confirmation of Evolutionary Theory.* 2nd ed. Princeton, N.J.: Princeton University Press.

Matthewson, John, and Michael Weisberg. 2009. "The Structure of Tradeoffs in Model Building." *Synthese* 170: 169–90.

May, Robert M. 1973. "On Relationships among Various Types of Population Models." *American Naturalist* 107 (953): 46–57.

May, Robert M. 2001. *Stability and Complexity in Model Ecosystems.* Princeton, N.J.: Princeton University Press.

Orzack, Steven Hecht, and Elliott Sober. 1993. "A Critical Assessment of Levins's 'The Strategy of Model Building in Population Biology' (1966)." *Quarterly Review of Biology* 68: 533–46.

Rademacher, Christine, Christian Neuert, Volker Grundmann, Christian Wissel, and Volker Grimm. 2004. "Reconstructing Spatiotemporal Dynamics of Central European Natural Beech Forests: The Rule-Based Forest Model BEFORE." *Forest Ecology and Management* 194: 349–68.

Roughgarden, Joan. 1979. *Theory of Population Genetics and Evolutionary Ecology: An Introduction.* New York: Macmillan.

Suppes, Patrick. 1962. "Models of Data." In *Logic, Methodology, and Philosophy of Science: Proceedings of the 1960 International Congress,* edited by E. Nagel, P. Suppes, and A. Tarski, 252–61. Stanford, Calif.: Stanford University Press.

Turner, Monica G. 1993. Review of *The Mosaic-Cycle Concept of Ecosystems,* edited by Hermann Remmert (1991). *Journal of Vegetation Science* 4: 575–76.

Tversky, Amos. 1977. "Features of Similarity." *Psychological Review* 84: 327–52.

Van Fraassen, Bas C. 1980. *The Scientific Image.* Oxford: Oxford University Press.

Volterra, Vito. 1926. "Variazioni e fluttuazioni del numero d'individui in specie animali conviventi." *Memorie Della R. Accademia Nazionale Dei Lincei* 2: 5–112.

Weisberg, Michael. 2006a. "Forty Years of 'The Strategy': Levins on Model Building and Idealization." *Biology and Philosophy* 21: 623–45.

Weisberg, Michael. 2006b. "Robustness Analysis." *Philosophy of Science* 73: 730–42.

Weisberg, Michael. 2007. "Who Is a Modeler?" *British Journal for the Philosophy of Science* 58: 207–33.

Weisberg, Michael. 2013. *Simulation and Similarity: Using Models to Understand the World.* New York: Oxford University Press.

Weisberg, Michael, and K. Reisman. 2008. "The Robust Volterra Principle." *Philosophy of Science* 75: 106–31.

Wimsatt, William C. 1981. "Robustness, Reliability, and Overdetermination." In *Scientific Inquiry and the Social Sciences,* edited by M. Brewer and B. Collins, 124–63. San Francisco: Jossey-Bass.

Wimsatt, William C. 1987. False Models as Means to Truer Theories." In *Neutral Models in Biology,* edited by M. Nitecki and A. Hoffmann, 23–55. Oxford: Oxford University Press.

Wimsatt, William C. 2007. *Re-engineering Philosophy for Limited Beings.* Cambridge, Mass.: Harvard University Press.

Wissel, Christian. 1992. "Modelling the Mosaic Cycle of a Middle European Beech Forest." *Ecological Modelling* 63: 29–43.

Woodward, Jim. 2006. "Some Varieties of Robustness." *Journal of Economic Methodology* 13: 219–40.

SYMPOSIUM ON MEASUREMENT

10

INTRODUCTION
The Changing Debates about Measurement

BAS C. VAN FRAASSEN

MEASUREMENT IS THE MEANS through which data, and subsequently a data model, are obtained. To understand what sort of data model can be obtained we need to understand what sort of data can be obtained. What sort of activities qualify as measurement, what do or can we learn through measurement, what are the limits of and constraints on what we can learn?

This subject has been the scene of philosophical debate for about a century and a half. Paul Teller takes as its background a simple, arguably naïve, realist conception of nature characterized by certain quantities and of epistemic access to the values of those quantities. Philosophical dismay with that conception began in the nineteenth century and drove much of the effort to understand how measurement relates to theory in science.

Two approaches to this problem dominated the discussion in the twentieth century: the *representational theory of measurement* and the considerably more realist *analytic theory of measurement*. Neither was aptly named. The former term is due to Luce, Narens, and Suppes, with reference to mathematical representation theorems (Luce and Narens 1994; Luce and Suppes 2002). The latter position, which postulates the sort of realism Teller castigates, was presented by Domotor and Batitsky (2008). However, both approaches are recognizable in a large diversity of writings. Teller's presentation of his pragmatist approach begins with a sharp critique that applies to both.

How did it all begin? A crucial new philosophical perplexity about measurement came with the shock to the theory of space, when it seemed that measurement would have to decide between Euclidean and non-Euclidean geometry. First reactions, by Gauss and Lobachevski, assumed that the decision could be made by measuring the interior angles of a triangle formed by light rays. But why must light ray paths mark straight lines? And what measures the angles? Straightness and, more generally, congruity are familiar as geometric mathematical concepts, but now they must be identified as empirical physical quantities.

Helmholtz's famous lecture of 1870 on physical congruence cast a bright new light on the ambiguities in what counts as, or indeed what *is,* a measurement of a particular quantity (Helmholz 1870/1956). Mathematically, a space may admit many congruence relations, hence the question whether physical space is Euclidean will hinge on which procedures count as establishing congruence between distant bodies. With several striking thought experiments Helmholtz argues that the same measurement procedures will yield alternative conclusions about the structure of space when understood in the light of different assumptions, not independently testable, about what is physically involved in those procedures.[1]

One of Helmholz's concluding remarks set the stage for a century-long effort to ground the theory of measurement in a conception of simple physical operations that would serve to define qualitative, comparative, and quantitative scales: "In conclusion, I would again urge that the axioms of geometry are not propositions pertaining only to the pure doctrine of space. They are concerned with quantity. We can speak of quantities only when we know of some way by which we can compare, divide and measure them" (1870/1956, 664).

Through a succession of writings on the subject in the century following Helmholz's lecture, a general theory of measurement developed (see especially Diez 1997a, 1997b). The early stages appeared at the hands of Helmholz, Hölder, Campbell, and Stevens. However, this project was effected most fully in collaboration with theoretical psychologists by Patrick Suppes in the 1950s and 1960s (Suppes 1969; Suppes et al. 1989).

This general theory, there called the *representational theory of measurement,* focused on the representation of physical relations in numerical scales. It was remarkable for two features: the sophistication of its mathematical development and the paucity of its empirical basis. In typical format, the results show that a domain on which a certain operation ("combining") and relation ("ordering") are defined can be, if certain axioms hold for that domain, uniquely represented by a numerical structure with + and <. The gloss on this result is that there is a quantity pertaining to the objects in the domain whose values are the numbers assigned in that representation. But on the side of the domains in question, to characterize those operations and relations we are offered *counterfactual conditionals* about manipulation. (For example, stick A is to be assigned a greater length than stick B exactly if, were they to be laid side by side, stick A would extend beyond stick B.)

The assumptions include ones that go beyond the finite, such as the "Archimedean" assumption about how any greater value of the quantity length would be surpassed by some finite amount of combining. As is honestly noted, to satisfy the axioms in question, those counterfactuals, even taken at face value, are entirely unrealistic, requiring infinitely fine discrimination (see, e.g., Batitsky 1998). This critique is strongly augmented by Paul Teller's arguments.

In the scientific theories that were kept in mind, the entities dealt with are characterized by quantities that have definite values. A body in Newtonian mechanics has a mass, position, and speed; each of these has a real number as its value. It has a direction of motion, a velocity, and an acceleration; each of these has a vector over the real number continuum as its value. The values change, but at each instant each of these quantities has one of its possible values. These quantities correspond to the dimensions of a logical space, and a *measurement* will locate a body in that space. So far, the representational theory of measurement was true in its aim and arrived at its proper target. But all this comes with a blank space about what a measurement is—what sorts of operations, under what conditions, count as measurement, or as measurement of a given theoretically defined quantity.

What is striking about this approach, in retrospect, is how the crucial point about the theory-dependence of measurement, which we already see coming to the fore in Helmholtz's 1870 lecture, was lost from sight. The focus in recent decades on scientific practice shifted attention to the question of how measurement, in the role of data-generating procedure, appears in a model of the experiment (e.g., Chang 2007). There is no escaping the fact that the model of the experiment with which the experimenter is working is itself a theoretical model. Whether a given procedure is actually generating data for the experimenter depends on whether this procedure itself can be theoretically represented as properly related to a quantity by which the theory represents features of the entity investigated.

We sense a threatening, enfeebling relativism or skepticism that looms in these reflections. "*Il n'y a pas de hors-texte*"[2] is a disturbing thought in any context, and the disturbing thought of vicious circles or "theoretical nepotism" has certainly been raised about these new thoughts about measurement. So it is not surprising to find a strongly realist reaction.

One response, ably presented by Zoltan Domotor and Vadim Batitsky, is to reject the demand for anything resembling an "operationalist" basis and to take a solidly realist line. That is, terms purporting to refer to quantities are theoretical terms, and to take a theory to be true is to take those terms to refer to real, objective characteristics present in nature. These characteristics and relations between them are there to be discovered, not invented; a measurement is a physical operation that evaluates a quantity, by means of interaction, to evoke a manifestation of its value. The theory of measurement, conceived in this form, is not a philosophical elucidation; it is itself a scientific theory of great generality, the theory of physical quantities.

Again, this rival account is developed with exemplary mathematical sophistication. The theory of measurement operations and measurement instruments developed and the innovative algebraic treatment of relations among quantities are valuable in their own right.

But to simply replace the naïve empiricist account with a postulational approach looks a little like just declaring victory and going home. That the gap between theory and nature can be bridged by a simple postulation of physical quantities that match their mathematized counterparts was roundly castigated already by Hans Reichenbach (cf. van Fraassen 2008, 118–121, 240–244).

There is a third way, which divides the labor between mathematical articulation and historical appreciation of scientific practice. We can think of that too as an empiricist way—but improving upon the naïve empiricism of the original motivation for the representational theory of measurement. In effect there are two points of view to adopt when studying any specific measurement (measurement of length, of mass, of temperature, of force, and so on). One view, which is very theoretical—"from above," so to speak—studies the measurement interaction as it is represented in the relevant theory. This study will include a particularization within a mathematical framework such as Suppes's or Domotor and Batitsky's. The other, which is predominantly historical and "from within" the relevant practice, studies how the theoretical concepts, models, and measuring operations evolved in mutual interaction. Neither will pretend to proceed from a "hygienic" theory-free basis; together they elucidate measurement as a situated activity, never outside a theory-laden and perspective-determined context.

The latter approach, exemplified in case studies that are at once philosophical and historical, was inspired by a take on the theory-dependence of observation and measurement that is quite different from the "realist" analytic theory of measurement. The seminal writings of Sellars (1948), Feyerabend (1957), and Kuhn (1962) demonstrated that the language in which observation and measurement results are reported is thoroughly and irremediably theory laden. There are always rival theoretical contexts, and the same "readings" take on different meanings; the same operations may have contrary significance or no significance at all, and they may not count as measurements of the same quantity or not count as measurement at all in these different contexts.

Teller explores here the repercussions of this view when conjoined with an appreciation of how highly idealized those theories are and how tenuous is any claim to accuracy (let alone truth) of even the best theories available. Not only accuracy and truth but reference itself becomes a challenge. Ostensibly scientific descriptions refer to concrete entities and quantities in nature, but what are the referents if those descriptions can only be understood within their theoretical context? A naïve assumption of truth of the theory that supplies or constitutes the context might make this question moot, but that is exactly the attitude Teller takes out of play. Theory laden, one might say, but laden with false theories! In retrospect, as we see when Teller pursues this theme, even the views of those seminal writers involved conceptions of science far removed from its actual practice.

Once the topic of measurement is approached in working context, immersed in experimental and modeling practice, very different if equally disturbing questions appear. These are the questions addressed by Paul Teller, who shows us the great distance between simplistic philosophical conceptions and the problems practically and pragmatically faced in scientific practice.

NOTES

1. For a longer discussion of his thought experiments and their philosophical impact, see van Fraassen (2008), 214, 229.

2. "There is no such thing as [what is] outside the text," the (in)famous dictum of Jacques Derrida, from *Of Grammatology*, trans. Gayatri Chakravorty Spivak (1967; Baltimore: Johns Hopkins University Press, 1976), pp. 158–59.

REFERENCES

Batitsky, Vadim. 1998. "Empiricism and the Myth of Fundamental Measurement." *Synthese* 116: 51–73.

Chang, Hasok. 2007. *Inventing Temperature: Measurement and Scientific Progress*. Oxford: Oxford University Press.

Diez, José A. 1997a. "A Hundred Years of Numbers. An Historical Introduction to Measurement Theory 1887–1990. Part I." *Studies in the History and Philosophy of Science* 28 (1): 167–85.

Diez, José A. 1997b. "A Hundred Years of Numbers. An Historical Introduction to Measurement Theory 1887–1990. Part II." *Studies in the History and Philosophy of Science* 28 (2): 237–65.

Domotor, Zoltan, and Vadim Batitsky. 2008. "The Analytic versus Representational Theory of Measurement: A Philosophy of Science Perspective." *Measurement Science Review* 8: 129–46.

Feyerabend, Paul K. 1957. "An Attempt at a Realistic Interpretation of Experience." *Proceedings of the Aristotelian Society* 58: 143–57.

Helmholtz, Hermann von. (1870) 1956. "On the Origin and Significance of Geometrical Axioms." Translated by J. R. Newman. In *The World of Mathematics*, vol. 1, edited by James Newman, 647–68. New York: Simon and Schuster.

Kuhn, Thomas S. 1962. *The Structure of Scientific Revolutions*. International Encyclopedia of Unified Science, vol. 2, no. 2. Chicago: University of Chicago Press.

Luce, R. Duncan, and Louis Narens. 1994. "Fifteen Problems Concerning the Representational Theory of Measurement." In *Patrick Suppes: Scientific Philosopher*, vol. 2, edited by Paul Humphreys, 219–49. Dordrecht, the Netherlands: Kluwer Academic.

Luce, R. Duncan, and Patrick Suppes. 2002. "Representational Measurement Theory." In *Stevens' Handbook of Experimental Psychology*, vol. 4, edited by J. Wixted and H. Pashler, 1–41. New York: Wiley.

Sellars, Wilfrid. 1948. "Concepts as Involving Laws and Inconceivable without Them." *Philosophy of Science* 15: 287–315.

Suppes, Patrick. 1969. *Studies in the Methodology and Foundations of Science*. Boston: Reidel.

Suppes, Patrick, David M. Krantz, R. Duncan Luce, and Amos Tversky. 1989. *Foundations of Measurement*, 2 vols. New York: Academic Press.

Van Fraassen, Bas C. 2008. *Scientific Representation: Paradoxes of Perspective*. Oxford: Oxford University Press.

11

MEASUREMENT ACCURACY REALISM

PAUL TELLER

YOU MEASURE THE TEMPERATURE of a glass of water and say that the outcome is accurate—is correct—to within a tenth of a degree. What does this mean? Presumably that there is some number that is, say, the temperature of the water in degrees centigrade, and that the measurement outcome is within one-tenth of a degree of that true value. The present discussion will work to undermine this supposition, though at the very end I will present a way of understanding such statements that is consistent with all the difficulties that will have come before.

Restrictions

I will restrict attention to physical quantities, though most of what I say should apply, with suitable modifications, to both the life and social sciences. I will also restrict attention to quantities, such as mass and temperature, that can be represented with a measurement scale of real numbers, as opposed, for example, to curvature which requires a tensor. But what I discuss explicitly should apply also to such multivariable quantities.

Initial Characterization of Measurement Accuracy

To fix on our target, we need to review some basics. First, we distinguish between measurement indications and measurement outcomes: an indication is "what is shown on the meter." But often such an indication can be corrected on a theoretical basis. A measurement outcome is the final result after such interpretation. Throughout I will have measurement outcomes in mind.

One can attribute accuracy to any measurement indications, outcomes, the instruments used to produce indications, or the entire measurement system comprised by the instrument and the theoretical basis used for interpretation. Although much of what I will have to say will apply to all of these, we are best off, again, taking measurement outcomes as our primary target.

Accuracy must be distinguished from precision. The standard analogy refers to arrows shot at a target. The outcomes are accurate to the extent that they are close to the bull's-eye. They are precise to the extent to which they cluster closely together. So measurement can be extremely precise without being very accurate.[1]

I take the default understanding of measurement-accuracy to be what I will call "traditional measurement accuracy realism." One supposes that there are in nature things such as lumps of lead and glasses of water, kinds of things such as lead and water, and quantities that pertain to things and kinds such as mass, length, temperature, and time (pertaining to duration of processes); and one supposes that in concrete cases such quantities have values. Stated generally:

Traditional measurement accuracy realism (stated schematically for measurement of quantity Q, with possible values q, in units u, on an object or type of object O)[2]:

Presupposition: There is in nature the quantity Q, with value q, in units u, for object or type of object O.

Then q', a measurement outcome of Q in units u on O, counts as
a) Perfectly accurate: $q' = q$.
b) Accurate (enough): the outcome, q', is close enough to q for present purposes.
c) Outcome q' is more accurate than outcome q'': q' is closer to q than is q''.

Accuracy understood in the traditional way is supposed to be an objective, not an epistemic matter. Realists will agree that accuracy can be estimated but not exactly known, but they insist that there is nonetheless a fact of the matter, just how accurate, in the traditional sense, a given measurement outcome is.

PROBLEMS WITH TRADITIONAL MEASUREMENT ACCURACY REALISM

General Statement of the Problem

Traditional measurement accuracy realism fails because the terms used in the relevant statement instances fail to refer. We use terms for quantities and their values: "The temperature of the water in this glass." Traditional measurement accuracy realism supposes that there is "in nature" some determinate quantity, temperature, or more specifically, the temperature of the water in this glass, that in this instance has some determinate value, say 20.258743 . . . °C. My claim is that the term "the temperature of the water in this glass" does not have a referent. My reason is not in any way metaphysical. It is simply that the full facts of language use and circumstances of utterance fail to pick out any one thing to be the named quantity "temperature," or any one number to be the claimed value of the claimed quantity. We will see a complex of detailed reasons for this failure, but at bottom they are all consequences of the contingent circumstance that the world is far too complex for our language to get attached to completely determinate things, in particular quantities and their value instances.

I must dwell on the form of my complaint because it is entirely different from what one usually hears from those known as antirealists, and my argument will be misunderstood if the reader falls back into thinking that I am attacking conventional realism in a familiar way.[3] I am not claiming that there are no quantities with exact values in nature, nor, as some antirealists would have it, that the whole idea of "things in nature" is incoherent.[4] Indeed, there is no coherence problem in such statements because we can model what this would be like.

Rather, to repeat this for emphasis, the problem is one of reference failure. Such determinate quantities as there may be fail to get attached to quantity terms, such as "time," "mass," "length," "velocity," or "temperature." With no determinate quantities attached to such terms, there are no determinate values for them to have. In addition, even if we suppose that the quantity terms do refer, we will see that determinate reference for terms purportedly referring to their values would fail anyway. We will also see difficulties with reference for terms for units, such as "kilogram," "meter," and "second."

The problem is also not epistemic in the sense that presupposes that our terms for quantities and their values do refer, but that there are problems in knowing just what those values are. Rather the claim is failure of the presupposition, that the relevant terms have been successfully attached to determinate referents.

One immediate reaction is to say, well, there are no point-valued referents, but we can always make do with an interval. But how is this interval to be understood? What one always has in mind is that the true value lies somewhere in the interval. But that takes us back to the questioned exact valued referents. I will examine questions about intervals in more detail later.

Reference Failure Source Points

There are different kinds of problems for three different kinds of what I will call "reference failure source points." The first is composed of quantities in the sense of a dimension as used in dimensional analysis. Mass, length, and time are usually taken as fundamental, and they figure in the characterization of other quantities such as velocity, which has the dimensions of length divided by time. I will refer to these collectively as "dimensional quantities." Dimensional quantities are theoretically individuated—that is, identified by the role that they play in our theories.[5]

Our next reference failure source point is the units used in characterizing a quantity. Without determinate units, no determinate quantity can have been picked out. Even if we had succeeded in specifying some quantity, say one called "mass," just what quantity is in question is still open until we have said whether it is mass in kilograms, in grams, or some other unit. When traditional measurement accuracy realists postulate an independently existing value for a quantity of an object on an occasion, where objective accuracy is some measure of the difference between this and a measurement outcome, the independently existing value and the measurement outcome must be understood in terms of the same units.

Finally, I will need to distinguish between dimensional quantities and what I will call "working quantities." Velocity is something abstract: Velocity of what? Velocity, or its absolute value speed, of sound in air is relatively speaking concrete; and speed of sound in air and speed of water in a pipe are different concretizations of the abstract speed.

One usually does not distinguish between the abstract dimensional and the, relatively speaking, concrete working quantities. In particular, metrologists appear to refer indifferently to dimensional and working quantities as measurands, for example, VIM 2.3 (BIPM 2012b).[6]

Measurand: quantity intended to be measured.

But the distinction does tacitly occur in both VIM and GUM (BIPM 2012a). From VIM 0.1:

> Even the most refined measurement cannot reduce the interval [that can reasonably be attributed to the measurand] to a single value because of the finite amount of detail in the definition of a measurand.

If one has dimensional quantities in mind, this statement puzzles because of the absence of any concrete mention of refinement of "definitions" of dimensional quantities. However, we see what is in question in GUM. GUM echoes VIM with

> D.1.1: The first step in making a measurement is to specify the measurand—the quantity to be measured; the measurand cannot be specified by a value but only by a description of a quantity. However, in principle, a measurand cannot be *completely* described without an infinite amount of information.

What is in question becomes clear with the following example, D.1.2:

> Commonly, the definition of a measurand specifies certain physical states and conditions.

> EXAMPLE The velocity of sound in dry air of composition (mole fraction) $N_2 = 0.7808$, $O_2 = 0.2095$, $Ar = 0.00935$, and $CO_2 = 0.00035$ at the temperature $T = 273.15$ K and pressure $p = 101,325$ Pa.

What is the infinite amount of information here referenced? Conceivably there is an indefinitely long list of such potentially relevant characteristics. But more likely it is the interval left open by all such specifications. It is understood that temperature is being specified as $T = 273.15$ K ± 0.005 K, and so on.

In any case, I need the distinction between abstract dimensional and (relatively) concrete working quantities because there are vastly different problems that arise for the two.

When working quantities are in question there will be some differences between type and completely concrete token cases. When discussing the speed of sound in air or the melting point of lead one has in mind the characterization of a property of a *kind* of substance—air or lead as a type. But one also needs to measure quantities for concrete instances—tokens—such as the speed of sound in the air in the Sydney Opera House at some specified time, or the temperature of the water in some specified glass at a specified time.

Difficulties with Working Quantities

As relatively concrete realizations of dimensional quantities, whatever problems will arise for dimensional quantities will, ipso facto, apply as problems for their concrete realizations. But working quantities present additional difficulties. Roughly speaking, these difficulties arise in either how their dimensional abstractions are made concrete or from the fact that they are not made completely concrete. To make these additional difficulties clear, for the discussion of working quantities we will take their dimensional abstractions as given and unproblematic.

Taking token cases first, consider a measurement of the speed of sound in the air in the Sydney Opera House at 8:00 p.m. on January 1, 2013. There are two difficulties. First, just what will we count as part of the Opera House? Include the vestibule? Oh, you will protest, obviously what was intended was the auditorium of the Opera House—but to no avail. With the door open or shut? Filled with an audience or empty? Any specification of a concrete object will leave open to some extent precisely what object is in question. Having failed to designate a determinate concrete object, there can be no determinate value that "it" actually has.

Second, the speed of sound will vary from one part of the Opera House (or the auditorium of the Opera House, or the . . .) to another. For example, speed of sound varies with temperature, and the temperature will not be absolutely constant throughout. And there will be edge effects.

Turning to type cases for working quantities, this is the problem from VIM and GUM quoted earlier. The problem could be understood in two ways. First, "speed of sound in air" is open ended, as is "speed of sound in air at temperature T = 273.15 K," and likewise "speed of sound in air at temperature $T = 273.15$ K and pressure $p = 101,325$ Pa." Could this list be continued indefinitely with more and more relevant features? Possibly, but that is a bit implausible, so we will let it pass.

But, second, how are the specifications to be understood? As mentioned earlier, most plausibly with a temperature of ±0.005 K and pressure ±0.5 Pa, the values in the intervals will give rise to different speeds of sound.[7] One could, on the other hand, take the specific characteristics of temperature and pressure to be intended as completely precise. But no real world sample of air has such precise values, if only because the values would vary slightly from place to place. So at best one is talking about the speed of sound in air . . . in some idealized condition, not in the real world.

Units

The characterization of units presents a whole new raft of problems. Except for the kilogram, fundamental units are now defined using a theoretical definition. For example, currently

> the second is the duration of 9,192,631,770 periods of the radiation corresponding to the transition between the two hyperfine levels of the ground state of the cesium 133 atom . . . This definition refers to a cesium atom at rest at a temperature of 0 K.[8]

This definition involves a number of idealizations.[9] Before getting specific I need to separate out the kind of problems that will be in question for us.

To operate as a standard such an idealized theoretical definition has to be realized in some concrete piece of apparatus that will in practice function as the standard, and so doing involves deidealization from the theoretical definition. One first constructs the needed apparatus so as to minimize as far as possible the departure from the idealized definition, and one then further deidealizes using theory-based adjustment of the indications physically produced by such instruments.[10]

This need for practical deidealization in physical realization of a standard differs from the implication of idealization that we will now consider. The practical case concerns the operation of some concrete device. In examining traditional measurement accuracy realism we are concerned with, rather, whether the theoretical definition succeeds in picking out a referent, picking out some real world characteristic, quite independently of the question of whether that characteristic can in practice be exactly realized.

The form of the problem is that the idealizations involved in a definition of a unit mean that the definition is of a unit in an idealized situation, speaking metaphorically, in a nonactual "possible world." There is no guarantee that what is picked out for one or more such nonactual possible worlds will correspond in the way needed to any one determinate referent in the real world. Examination of cases shows that this is exactly what is in question.[11]

Let us consider first the one unit that is still "defined" by a physical standard, the kilogram characterized in terms of the international prototype kilogram. Taking this as a perfectly precise characterization of what mass will count as a kilogram involves idealizing away variable factors, such as contaminants from the air and scratches induced when the prototype is handled while making replicas, both problems that managers struggle to minimize but can never completely eliminate. Strictly speaking, sublimation of the material of which the prototype is composed has also to be idealized away. Or, if one refrains from such idealization, there is no one mass that the prototype picks out over time because the complications such as the ones just mentioned mean that the mass of the prototype varies up and down.

Even at one time there is no completely determinate real world mass that is picked out—for the same reason that gave rise to one of the complications for token cases of working quantities. Absolutely precisely, just what is, even at a fixed time, *the* prototype? No one answer to this question will pick out an object that will provide the kind of standard that we assume. A policy either of including or of not including the present scratches picks out, at best, a standard that will be different as soon as the prototype is handled. Or to give a circumstance that is utterly inconsequential in practice, strictly speaking relative motion of exactly zero is an idealization. Real world uses will involve relative motion and thus an indeterminacy in what is in question: rest or relativistic mass, and if the latter, which one? Although this is utterly inconsequential in practice, the realist requires a *completely precise* value.

Other standards are defined theoretically. Consider the theoretical definition of the second, as mentioned previously. This definition ignores the time–energy uncertainty relation that results in spectrum bandwidth. Given the bandwidth, the definition does not pick out any unique real world temporal duration. Or again, appeal to a temperature of 0 K. Nothing in the real world can be at 0 K, nor can 0 K be approached asymptotically because of the finite limit imposed by quantum vacuum fluctuations. At best the definition characterizes a temporal duration in some possible world. In fact in many possible worlds because there is no unique way in which the idealizations can be removed. (What will a possible world with no quantum effects be like?) There will be no sense to be made of which of such possible worlds is "closest" to the actual world, so an appeal to "closest world" will not pick out a unique real world temporal interval.

And we are not done. The general theory of relativity and quantum field theory used in the theoretical definition are themselves idealizations—two theories that are not unified, and of which it is at least questionable whether they are mutually consistent. These idealizations provide further reasons why, strictly speaking, the definition only gives a temporal interval in some, or really in many possible worlds.

The definition of the meter also fails to deliver the completely determinate length that realists require. Bureau International des Poids et Mesures gives the definition of the meter as "The meter is the length of the path travelled by light in vacuo during a time interval of 1/299 792 458 of a second."[12] This definition inherits all the problems of the definition of the second. It involves the further idealization of the speed of light in vacuo, and the idealization of general relativity applies anew, now through its ideal treatment of distance.

Dimensional Quantities

I have saved the most vexing case for last: the case of dimensional quantities. As I mentioned, dimensional quantities are individuated by the theories in which variables for these quantities occur. But the theories in question are all idealized. So in the real world there are no quantities as characterized in our idealized theories. If they occur anywhere, it will be (again, speaking metaphorically) in the idealized possible worlds of the characterizing theories.

Take the example of mass. Is this supposed to be Newtonian mass? Relativistic mass? The mass of quantum field theory that is a renormalized quantity and thus dependent on the "impact parameter" involved in its measurement? Quantum field theory is still highly idealized, so there is good reason to think that further deidealization will further recharacterize just what quantity is in question.

One wants to protest—these increasingly accurate characterizations are all of one quantity of which our theories are giving an increasingly faithful account. I will discuss the "close to" worry in a general way later. But the example of mass helps to make clear the weakness of the response. The mass of quantum field theory is so different from that of Newton that the idea that we are just refining an already very clear idea looses all plausibility. The *only* constraint on further deidealization is that old successes be preserved. These old successes may be preserved by radically new ideas of quantities. This can happen by the operation of a limit. In the relation between special relativity and Newtonian mechanics, one gets the latter from the former by letting v/c go to zero. But that does not make the Lorentzian metric and its geometry just a refinement of Euclidean geometry.

Let us try time. Our best theory of time is the general theory of relativity (GTR). But GTR is not quantized, and current efforts to quantize GTR play havoc with the treatment of time. We do not know the outcome of this story, but at the very least there is the lively possibility that a better theory characterizing time may characterize it as differently from GTR as quantum field theory characterizes mass as compared to Newtonian or relativistic mass.

Let us try another quantity, velocity. Velocity does not occur as a quantity in quantum theories. When we can ignore quantum corrections, one takes *speed* (magnitude of velocity) to be the limit of average speed. But the limit of averages is another idealization, one that breaks down badly even before we get to quantum corrections. And if by speed we mean an average speed, which average?

What about length? When one takes into account the indeterminateness of relative position as characterized in quantum theories, there is no such quantity. Indeed, in quantum theories length, or (relative) position, is characterized as an operator not as a real valued quantity—again, a radical departure from prior conceptions. Likewise in quantum theories momentum is a radically different kind of quantity from prior classical characterizations, like quantum mechanical position also characterized by an operator, not by a real number. These few words paper over a great many complications but should be enough to show that there are serious issues for the case of both position and momentum.

Repair by Appeal to Intervals?

The realist in us all is screaming. True, no objectively occurring precise values are attached to our terms. But, objectively, suitable intervals (or other collections of values) can do the needed realist work. Here I consider this option, construed in terms of completely determinate collections of values— that is, collections for which, for each number, there is a fact of the matter whether it is in the collection or not. Later I will consider "indeterminate collections" (starting with the question of what that could even mean).

How should such an interval be understood? What one wants to say is that we are talking about an interval of values that are, in some sense, "close enough." But close enough to what? For realism, as we have construed it, in a given problem situation there must *be* a value the closeness to which counts as "close enough," however that is to be understood. But for all the reasons previously given, there is nothing in the problem situation that fixes the needed objective value.

For the case of working quantities there is a more careful way to make out the interval intuition. Let us see how this goes for the speed of sound in air. To review the problem, specifying a quantity as "speed of sound in air" is, as VIM and GUM would put it, an incomplete definition. Liquefied air? Ionized air? It is plausible that all such extreme cases can be eliminated with a short list of more specific conditions: air at temperature $T = 273.15$ K, and pressure $p = 101,325$ Pa. But such characterizations of the quantity are still open ended: in the present example, temperature ±0.005 K and pressure ±0.5 Pa. The proposed solution, in the spirit of supervaluationism, suggests that we get our interval by considering *all* the ways in which the characterization could be made completely precise. To put it once more metaphorically, consider the possible worlds each having some precise value for the quantities in question (temperature, pressure), the range of possible worlds fixed by the limits in such incomplete specification of the quantities and that are otherwise maximally similar to the actual world. Our required interval (or other collection) of values will be the values in one or another of such possible worlds.

Such an interval would be objective. The statement of realist accuracy would have to be restated: instead of distance from some one value there would have to be some relation to the interval of question. This could be done in a variety of ways, the details do not matter.

This proposal collapses, in different ways, depending on how velocity is understood. Let us suppose, which is what one usually has in mind, that it is instantaneous velocity that is in question. Again, this is an idealization—there is no such thing in the real world. The proposal is to consider a range of possible worlds that differ from the real world *only* by having one or another precise value of the associated quantities, such as pressure and temperature, that are within the bounds of the interval specified in the detailed characterization of the condition of the air in which the speed of sound is in question. (For the moment we are waving aside the problems with both pressure and temperature, which, when reintroduced, will further spoil the effort.) But with these worlds differing from the real world only by variation of the exact parameter values within the given bounds, these possible worlds will also have no instantaneous velocities. If velocity means instantaneous velocity, the proposal is empty.

The alternative is to consider some kind of average velocity in each of the relevant possible worlds. But which? No question but that there are averages—distance covered divided by the time of travel—that will work for practical purposes. (I am taking the appeal to "practical purposes" to paper over the problems with the appeal to the distance and time of travel. This broaches problems of vagueness, to be discussed below.) But the realist needs to be specific. "Pick some average that works for our current objectives" does not fit the bill in the actual world, let alone in all of the various possible worlds relevant in the proposed analysis. In addition there are problems with the averages themselves. Wave or group velocity? Wave velocity is strictly defined only for a wave that extends to infinity forward and backward. And distance traveled in unit time brings in all the problems with measures of both distance and time—the problems we have already reviewed both for the units in question and for the more fundamental quantities of distance and time.

The interval intuition fails, if anything more radically, when it comes to units. At first things look hopeful because we are told, for example, that the current practical accuracy for standards for the second is to five parts in 10^{-15}. But what does this mean? As we will learn in more detail later, it means that concrete standard realizations can be built to agree to five parts in 10^{-15}. It is not yet clear what that shows about some kind of objective interval in nature. The agreement in practice clearly has some kind of controlling objective element inasmuch as nature makes us work very hard to get the agreement. But to what one thing "in nature," whether point valued or precise interval, in terms of which realist accuracy might be characterized, does this "objective element" correspond?

Unlike the case of working quantities, there is no natural candidate for the needed interval. For working quantities one plausibly turned to all the different ways in which an incomplete specification of the working quantity might be filled in. But in addressing the idealizations involved in the characterization of a unit there is no natural or well-defined range of cases of what will count as a deidealization. The only constraint on deidealization is that past successes be preserved, that in the case of units amounts to the successes in getting real-world realizations to agree at least as well as before any new deidealization. But what would be meant by the "interval of deidealizations" that might sustain the level of agreement that we now achieve in practice?

Dimensional quantities suffer, for this issue, from the same problems as do units. Because dimensional quantities are abstract, unlike their concretizations in working quantities, the whole idea of an interval of refinements has no direct application. As in the case for units, any idea of an interval would have to be in terms of some range of deidealizations from the idealizations involved in the characterization of the dimensional quantity in question. It is obscure in the extreme what kind of an interval could correspond to departures of our current idealization from one or another possible "finally correct" definition of a quantity. There would have to be some kind of objective distance measure between our current idealized definition and what a "final definition" might be. As in the case of units, the only current constraint on a "final definition" is that it preserve current successes. But in the case of dimensional quantities, creatures of fundamental theories, the success of a fundamental theory is entirely entangled with the work done by other theories, fundamental and nonfundamental. What would it mean to say that this success delimits some kind of "interval" or other collection of cases reflecting facts about nature?

HOW TO UNDERSTAND MEASUREMENT ACCURACY

What to Make of All These Considerations

For a variety of reasons, in any instance of measurement there is no completely specific value that is determined by the total situation that fixes the objective (though unknown) accuracy, in the sense of difference between some supposed actual value and the value that is the measurement outcome. We have considered and found wanting an effort to substitute some kind of interval or other collection of values for an objective value. Yet there is no denying that in any actual case of measurement there is a range of values that are, as a matter of objective fact, reasonable ones that could be used, and comparison with any of which gives a measure of accuracy. Note the shift, in the last sentence, to the epistemic notion of *reasonably assigned values*. These are still objective, inasmuch as there are a right and a wrong, or at least a more or less reasonable, that constrain what we should do and that indirectly reflect what is going on in a world too complicated for us to know exactly.

For a sensible idea of how this works, we should look at how metrologists evaluate accuracy.[13]

How Metrologists Evaluate Accuracy: Robustness Accuracy

As I have been at pains to emphasize, our understanding of quantities and how they might be measured is hostage to our currently best theories. Time is characterized by GTR, temperature by thermodynamics and statistical mechanics, and so on. Also central are theories that describe the interrelation of the quantity in question with other quantities. Where time is measured by periodic motion, crucial are theories of the motions in question. The current definition of the second appeals to a spectral emission of cesium, the theory of which calls on quantum field theory. Temperature is measured by the temperature dependence of other quantities such as the volumes of gases and liquids, the electrical properties of substances, and again the spectral properties of electromagnetic emissions for heated substances. Designing and evaluating measuring instruments for temperature requires applying the theories of these substances and the relation of temperature to their other properties.

Let us look in a little more detail at how this plays out in the case of determination of units. A unit, such as the second or the kilogram, is given a theoretical or physical "definition." The theoretical characterizations require various idealizations, such as 0 K and a zero gravitational potential. The physical prototype for the kilogram functions as a fixed standard only under idealizations such as no scratches when handled and no absorption of impurities. The theoretical characterizations then must be physically realized. While the prototype for the kilogram is already physically realized, the same problems that arise for physical realization of theoretically defined units arise for a physically defined unit in the form of the need to make copies. The physical realizations or copying depart from the idealized theoretical definitions and ideal circumstances assumed for a physical standard. To make effective use of a standard, one must, in physically realizing or copying, insofar as possible minimize these departures from the idealized definitions and conditions; and one further appeals to any relevant theory for help in further correcting for departures from the idealizations insofar as these departures still affect the physical realizations and copies. As we have seen, such deidealization cannot be done in any perfectly exact way, and what we come up with is hostage to the theories we use. Still, these theories are the best account of nature that we have, and we use them as best we can.

Metrologists work to keep track of such departures from the idealizations with what they call "uncertainty budgeting," which appeals to theory to estimate the uncertainties that arise as a result of failure to completely deidealize.[14] To be sure, these departures are not from something exactly fixed in nature but from standards that are as characterized by theories that are themselves idealized. That is, it is understood that these uncertainty estimates are relative to the theories used and thus limited by the shortcomings of these theories. In consistency with all of the worries elaborated above, these are not estimated departures from something fixed in nature but from the ideal depicted by what we take to be our best theories.

It is these estimated uncertainties, deployed in a robustness condition, that then provide the basis for attributing a level of accuracy to a measurement standard. In Tal's account of the special case of the standard second, one uses

> two interlocking lines of inquiry: on the one hand metrologists work to increase the level of detail with which they model clocks. On the other hand, clocks are continually compared to each other in light of their most recent theoretical and statistical models. The uncertainty budget associated with a standard is then considered sufficiently detailed if and only if these two lines of inquiry yield consistent results. The upshot of this method is that the uncertainty ascribed to a standard clock is deemed adequate if and only if the outcomes of that clock converge to those of other clocks within the uncertainties ascribed to each clock by appropriate models, where appropriateness is determined by the best currently available theoretical knowledge and data-analysis methods. (Tal 2011, 1091)

We have essentially the same story for the accuracy of measuring instruments proper. One provides a theoretical model for an instrument, relying on theory to minimize insofar as possible the uncertainties in the sense given here. Insofar as practicable, such models will take into consideration all the factors that, according to current theory, might affect the measurement process. One then uses these models to estimate the residual uncertainties, the inaccuracies to which the instrument might still be subject, once again according to our best theories. All the estimated uncertainties are combined, and combined with the overall uncertainty in the unit standard used in the calibration of the instrument.

The estimated uncertainties, deployed in a robustness condition, then provide the basis for attributing a level of accuracy to an instrument. Tal's summary is:

> Given multiple, sufficiently diverse processes that are used to measure the same quantity, the uncertainties ascribed to their outcomes are adequate if and only if
> (i) discrepancies among measurement outcomes fall within their ascribed uncertainties; and
> (ii) the ascribed uncertainties are derived from appropriate [as described previously] models of each measurement process. (Tal 2012, 175)

Uncertainties that satisfy this robustness condition qualify as reliable measures of the accuracies of the measurement outcomes of the instruments in question.

But Why Should Such Uncertainties Count as Measures of Accuracy?

One may take the robustness condition to proceed in the following spirit.[15] The world is too complicated for us to be able to describe it exactly as it is. We have to rely on a network of (not always exactly consistent) idealized theoretical accounts. But we use these accounts precisely because they give us a good enough picture to get along for a wide range of objectives. The uncertainties that figure in the robustness condition are not interpreted as uncertainties of departure from the realists' actually occurring values but as departures from values that we can suppose would occur in the idealized circumstances described by our theories.[16] Broadly, our composite idealized accounts are good enough to be highly reliable, and it is just a special case of this overall reliability that we will not get into trouble by treating departures from supposed idealized values of idealized quantities characterized in idealized units as departures from postulated actually occurring values of real quantities described in exactly characterized units. In the larger idealized picture of the subject matter, the measurement outcome is off by some (not exactly known) definite value from *what it would be in the (or some) simplified world* characterized by our idealized larger picture.[17]

Accuracy realism fails because of reference failure, and reference fails because of a fact that we too easily let drop out of view: the ubiquitous idealizations of our theoretical accounts of the world. We forget the idealized status of our theories precisely because they work so well and so broadly. Generally speaking, we get on successfully treating the world as characterized by the idealized dimensional quantities, specified in idealized units, and then applied more specifically with the idealized concrete versions provided by working quantities. In short, we proceed *as if* the presupposition of traditional measurement accuracy realism were true. In other words, the presupposition of accuracy realism is itself an additional idealization, or perhaps a collective application of prior idealizations.

Measurement standards function for us as the benchmarks against which measurement accuracy is evaluated. But that comes down to saying that we treat objects as having values for quantities as characterized in terms of our current measurement standards. On the one hand, we know that these standards are always susceptible to improvement, in ways in part marked by the ascribed uncertainties. But at any moment we can do no better than to treat the world as characterized in terms of these standards—that is, as if the world were just as so characterized. Acknowledging that improvement is always an option comes to acknowledging that using a standard as our guide to the world is an idealization.

The robustness condition is essential to the success of so proceeding. The condition functions as a prescription to check—with great thoroughness—that the various ways in which we assign values to quantities as described in our theories all fit together well enough not to engender difficulties.[18] The robustness condition functions precisely to ensure that taking the presupposition of accuracy realism, made concrete in terms of our measurement standards, as an additional idealization or collective application of antecedent idealizations does not spoil the larger operation of the sketch of the world provided by concrete application of our interconnected idealized models and theories.

The current proposal is not to scrap the concept of traditional accuracy realism in favor of some substantially different concept. Rather, I am urging a change in how we think about the concept. We apply the familiar concept, but no longer in a traditional realist spirit. Instead we appreciate its status as an idealization. Consider some specific measurement situation with an object of measurement being evaluated for the value of some quantity as characterized by our relevant current theories. Satisfaction of the robustness condition ensures that if one were to use any realization of the available measurement standards for this quantity one would get the same value up to the tolerances characterized by the uncertainty budgeting. Given this reliable consistency we will not get into trouble by idealizing, by thinking of the situation as one in which there *is* a quantity characterized by our theories, a unit set by the measurement standards, and that the object has a value for that quantity in those units. This last is just to say in the material mode exactly what is re-expressed in the formal mode by saying that this expression:

> There is a quantity characterized by our theories, a unit set by the measurement standards, and that the object has a value for that quantity in those units.

has precise referents, the quantity, the units, and the value in question. We know that these expressions do not have referents, but there is much practical advantage and no harm done by treating them as though they do.

As for the accuracy of some instrument that is not part of the system of measurement standards, we think of it within the scope of the idealization as the difference between the measurement result and the supposed actual value. Of course, even if the world were as in the idealization, the best we could do to get that supposed value would be the values of one or another measurement standard, qualified by the uncertainty budget. But just as in the real world the very best we could presently do would be exactly those results of one or another measurement standard, in practice what we can have is exactly what we would have if the world were as in the idealization. With the robustness condition in place, the idealization cannot get us into trouble.

QUANTITIES AND UNITS UNDERSTOOD AS VAGUE

Some readers will have been thinking throughout: the problems here are all problems having to do with vagueness. So there is no special problem here, nothing problematic over and above whatever general problems there may be with vagueness.

I agree, at least for the case of working quantities and units. The proposal, however, does not work for the case of dimensional quantities. In this section I will examine the connection and argue that treatment in terms of vagueness and in terms of idealization as in the last section are really two different ways of getting at the same thing. Certain advantages accrue to working with idealization, starting with the circumstance that framing in terms of idealization gives a uniform treatment of dimensional quantities along with units and working quantities, also thereby providing a kind of generalization of the notion of vagueness.

On its face, "accurate" is vague in exactly the way that "flat" is vague, but "accurate" is not the target there. Rather it is the expressions that have the form of picking out units, quantities, and their values. Compare:

> The temperature of the water in this glass.
> The time at which John arrived home.

There is no one temperature that counts as *the* temperature of the water in this glass. If you think that temperature is an intrinsic quality of objects, no one number will do—any real body of water in a glass will have some temperature gradients. Perhaps you want to take a statistical mechanical definition of temperature, the mean kinetic energy of all the molecules in the glass—but this too will suffer fluctuations and is, in any case, a classical idealization. Likewise there is no one precise moment that counts as *the* moment at which John arrived home. Was it when he pulled his car into the driveway? (And just which moment was that?). Or when he stepped over the threshold? Or when he hung his hat on the hat rack? "The time at which John arrived home" is vague, and in an analogous way so is "the temperature of the water in this glass."

Note that this proposal importantly differs from the interval proposal considered previously: there is no determinate interval of values that could count as the time of John's arrival. Likewise there is no determinate interval of values that could count as the water's temperature. Next we will consider how to make sense of a contrasting "indeterminate" collection of values.

If working quantity terms are vague, then there appears to be a simple way to characterize "accurate." Let us call the imprecisely characterized collection of values that would work for the temperature of the water in this glass the "temperature value collection." We could then characterize

Accurate (enough): Close enough for present purposes to any one (or almost any one) of the values in the temperature value collection.

Problem solved! For this analysis, "accurate" comes out as doubly vague: vague in the "enough" (compare: flat [enough]), but also vague in the "values in the temperature value collection."

We understand this approach precisely as well as we understand "temperature value collection." Above I rejected the option of saying that, though imprecisely characterized, there must *be* some determinate interval or other collection of values that is in question. We need to develop some alternative way of thinking about what is going on when we talk about such intervals or collections.

Let us work this through for "the time at which John arrived home." Suppose that we check the security camera and find that John turned off his car at 4:59:32, and crossed the threshold at 5:00:02; his hat hit the hook on the hat rack at 5:00:05. For virtually any practical purpose that might come up, you could use any of these times for "the time at which John arrived home" as well as many others that are, from a practical point of view, "close enough," though 5:00:00 would be the obvious practical choice. Which numbers could be used is open ended in the sense that which ones would be appropriate choices depends on what is at issue, in turn fixed by the context; however, "the context" itself will shift from case to case and in no case will be specific in every respect. At the margins one is free to cut the edges as one likes, and when the margins are fine enough choices will be arbitrary.

There is no determinate collection of values that qualify. There are only the practical questions of what numbers will serve, and how well, for practical issues. Yes, the approach illustrated in this example supplies an approach that could be applied very generally to the phenomenon of vagueness. Tal's robustness analysis is attractive because it provides a basis for making out this kind of thinking in an exceptionally general and coherent way for the case of measurement accuracy.

I do not know of any explicit development of this approach to vagueness in the vast vagueness literature. The usual way of dealing with the worry of indeterminately specified collections is the hierarchy of higher order borderline cases. Such developments may provide interesting formal constructions, but they are terrible models of vagueness of terms in natural languages and in particular in the languages of science. A borderline case is a case in which we are appropriately unsure about what to say. A borderline-borderline case would be one in which we are appropriately not sure whether we are appropriately unsure about what to say—something that in some cases we can make sense of, but something that in practice arises extremely rarely, if ever. The third order case goes beyond any normal human capacity or need, and so beyond anything that corresponds to the function of human language. The pragmatic approach to vagueness fits these conditions perfectly.

Understanding vagueness as a practical question of applicability makes it easy to see the connection with idealization. In the case of the temperature of the water in this glass, given the practical and theoretical questions of applicability that might come up we have some latitude as to which number to use for the temperature. The choice of any one is an idealized description of a much more messy real-world situation. An idealization is, strictly speaking, false. But within its domain of applicability one can use it as if the world were just as the idealization says it is. That is, for a suitable range of practical or theoretical questions the idealization functions as a precisification of a corresponding imprecisely characterized situation. In the case of the temperature of the water in our glass, postulating a precise value for the temperature is a precisification of the imprecisely characterized temperature value collection, one that is appropriate just when it is one of the values that arise in the practical analysis of the kind we have seen above. I call precisely stated idealizations and corresponding imprecise, or vague, characterizations "semantic alter egos" because they are different ways of accomplishing the same semantic work.[19]

But working with idealizations has the advantage that it will apply in the treatment of dimensional quantities where, I will now argue, thinking in terms of vagueness no longer applies.

To understand a vague term requires understanding how to make it more precise. This we can easily do for working quantities. For units I could make a case either way, depending on how we make more precise the vague "understanding how to make it more precise." But dimensional quantities cannot be forced into this mold for the kinds of reasons that already came up when we discussed the precise interval option for dimensional quantities. What would count as precisifications for terms for dimensional quantities would be deidealizations. But for our currently most detailed theoretical account of a dimensional quantity we have no idea how to deidealize—if we did, we would already have these proposed theories on the table!

To make this out in more detail requires addressing a complication. If the issue is put not in terms of an attribute of theory, idealization, but instead an attribute of language, vagueness, we also have to bring in the phenomenon of ambiguity because many of the relevant idealizations in question will correspond to ambiguity rather than vagueness. *Vagueness*—susceptibility to precisification from an indefinite range of refinements—and *ambiguity*—susceptibility to disambiguation from a determinate, very limited collection of determinate meanings—are not the same phenomenon.[20] For many theoretical considerations about language they must be distinguished. But for our purposes we can lump them together. The two share the relevant feature that to understand a vague/ambiguous term in a way that involves awareness of the vagueness/ambiguity requires knowing how to make the term more precise/knowing how to disambiguate the term. While not absolutely clear, it is at least odd and/or misleading to say that a term, as used in a language community, *is* vague or ambiguous even though no one in the community has any awareness of that vagueness or ambiguity or any understanding of how to precisify or disambiguate the term.

The kind of problem we are considering for dimensional quantities turns on variation in the discrete parameter of theory. Consider, for example, mass. As the term is now used it is ambiguous, between rest mass and relativistic mass. But this is only after 1905! I submit that as used before 1905 the term "mass" was not ambiguous—it referred to Newtonian mass, now best understood as rest mass. Before 1905 no one knew of the relativistic alternative, so disambiguation was not an option.[21]

The case for dimensional quantities differs from that of units and working quantities in two ways. First, it is ambiguity, not vagueness that is in question. From the perspective of our present interests this is an irrelevant difference. But second, when it comes to the dimensional quantities as characterized in our currently most detailed theory, the terms for dimensional quantities do not count as ambiguous. As noted previously, if we could disambiguate, this would be by appeal to more detailed theory that, in the cases in question, we do not have.

But the cases in which we can precisify/disambiguate and those in which we cannot still have in common the underlying source: idealization. We can appreciate that our characterization involves idealization. But we may or may not know how, at least to some extent, to deidealize. When we do know, we have vagueness or ambiguity; when we do not know, we do not. This is the reason for which our present subject is more perspicuously approached in terms of idealization rather than vagueness and ambiguity.

CONCLUDING THOUGHT

I have argued for the systematic failure of reference for referring terms for quantities and their values. We have seen this failure as a kind of generalized kind of vagueness (and ambiguity). Because vagueness is a ubiquitous aspect of language, in and out of science, this suggests that reference failure is likewise a very general feature of language. Braun and Sider claim just this, taking this circumstance to be sufficiently obvious that no argument is required: "The facts that determine meaning (for instance, facts about use, naturalness of properties, and causal relations between speakers and properties) do not determine a unique property to be the meaning of 'red' [and likewise for expressions very broadly]" (2007, 134). The earlier sections show in detail that this is so in the special case of terms for quantities, their units, and their values.

Just as the problem of reference failure—aka generalized vagueness/ambiguity—generalizes to all human representation, I urge that the refashioning of measurement accuracy by metrologists likewise generalizes: it is through idealizations that we know the world. The world is too complex for us to have representations that characterize it exactly—that is, with both perfect precision and perfect accuracy. Our representations always fall short in one or both of these two ways. This is as true of perceptual as of theoretical knowledge. But we do know a great deal. Knowing the world is knowing the world through idealizations, and insofar imperfectly.

It is a fair question: What is it to know the world through idealizations? This is a question on which I have touched in many other articles but which needs much more thought and discussion.[22] Indeed, it requires a wholesale overhaul of our understanding of human knowledge. For the moment I will leave it with the suggestion that the present treatment of measurement accuracy and its appeal to Tal's robustness condition provide an exemplar that can usefully guide our thinking.

The problem I examined at the beginning of the discussion is the semantic problem of reference failure, not an epistemic problem of difficulty in knowing values that are alleged to have been fixed. But the suggestion of the present section is that ubiquitous reference failure gives rise to a very different epistemic limitation, that we know the world only through idealizations. Philosophical tradition to the contrary not withstanding, knowing imperfectly is still knowing what the world is like; in particular, the world is very like one occupied by such and such idealized objects with such and such idealized characteristics. One does not have to get it exactly right about what things there are and their properties. There is a difference between getting things wrong in ways or to an extent that do not presently matter and getting things badly or completely wrong. Complete precision and accuracy is not humanly attainable and also not needed. Imperfect knowledge is still knowledge of the world—we can add redundantly, of the way the world is *really*. This is not traditional realism, but it is the sensible way in which we should have been understanding realism all along.

NOTES

1. I have written about measurement precision elsewhere (Teller 2013).

2. Reference to units is a short way to cover the point that what is postulated are not values as numbers, in some Platonic sense, but a ratio or other relation between the quantity Q, as it applies to O, and the quantity Q as it applies to some reference object (e.g., the international prototype kilogram) or condition (e.g., the radiation spectrum of cesium) that sets the units.

3. I do not consider the view I present in this discussion to be antirealist. Indeed, as I will explain in the last paragraph, the present view is the sensible way that realism should have been understood all along.

4. Nor would I claim that there *are* such things "in nature," whatever that might mean. This discussion is entirely agnostic about this question. If the reader *must* know what my private view is on this matter, let me just say that it is deeply Kantian.

5. As argued by Tal, who concludes that "in order to individuate quantities across measuring procedures, one has to determine whether the outcomes of different procedures can be *consistently modeled in terms of the same parameter in the background theory.* If the answer is 'yes,' then these procedures measure the same quantity *relative to those models*" (2012, 84, italics in original). This quotation also makes it clear that Tal is here referring to dimensional quantities, as opposed to what I will be calling working quantities.

6. I will be referring to two documents published by the Joint Committee for Guidelines in Metrology (JCGM): the *International Vocabulary of Metrology* (VIM) and *the Evaluation of Measurement Data—Guide to the Expression of Uncertainty in Measurement* (GUM). The references will be by section number (BIPM 2012a, 2012b).

7. If, with VIM and GUM, we take "definitions" of quantities to include such detailed specification of properties, air with these differing quantity values counts as different quantities. On this interpretation no specific quantity has been picked out.

8. "Time and Frequency: SI Unit of Time (Second)," Bureau International des Poids and Measures (BIPM), n.d., https://www.bipm.org/metrology/time-frequency/units.html.

9. I will use "idealization" very broadly to characterize any representational inaccuracy. This can but need not be understood as in comparison with some absolute standard. The alternative is to think of inaccuracy of a representation as what is so characterized from the point of view of some other representation that improves on the first in the sense of preserving all past and improving on the descriptive successes of the first representation. (Clearly in this note I am using "accurate" and "inaccurate" in a much broader sense than in the rest of the paper.)

10. See Tal (2011, 1088–90).

11. It is a sensible question whether the problems that the use of idealizations generate here are special to the case of measurement, quantities, units, and accuracy; or whether they are really more general. I urge the reader to put this question aside for the moment and focus on the argu-

ments. In the final section I will address the question of whether these problems are really more general.

12. "Metres, m," in "Base Units," Bureau International des Poids et Mesures (BIPM), n.d., https://www.bipm.org/en/measurement-units/base-units.html.

13. In this discussion I am closely following Tal (2011, and 2012, chapter 4).

14. I follow Tal (2011), especially pp. 1090–93.

15. Tal appears to have a different attitude toward his robustness condition. He never considers possible failure of the presupposition of what I am calling traditional measurement accuracy realism (see Tal 2011, 1094).

16. But note that, although they are not interpreted as uncertainties of departure from the realists' actually occurring values, they might also be that. Remember that this discussion does not argue that there are no potential referents in nature but that, should there be such, our terms fail to attach to them.

17. Remember that I am using "idealization" very broadly to characterize any representation that stands to be improved by being made more accurate in some way, where "improved" may be understood as relative to some descriptively more successful representation. See note 9 of this chapter.

18. Note that this way of thinking about the robustness condition differs from thinking of it as a regulative ideal of bringing language and reality into perfect alignment. I do not think that these attitudes exclude one another. It would be a very useful way to further explore these issues to consider the pros and cons of both these attitudes.

19. See my work for much more detail (2011 and 2017). Once the connection between idealization and vagueness is made, the arguments for reference failure in the early parts of the discussion function as an extended argument against epistemic accounts of vagueness.

20. See Bromberger 2012, 75–8 and passim.

21. A slightly more careful version of this account: prerelativistically, "mass" was ambiguous between inertial mass, gravitational mass, and the pretheoretic quantity of matter. Newtonian inertial mass was nonrelational, and in special relativity it is replaced by relativistic inertial mass that is relative to an inertial frame. This makes inertial mass, after 1905, ambiguous in a way similar to the ambiguity in "heaviness," ambiguous between the relational quantity, weight, and the nonrelational quantity, mass, a dimension of relationality that layers on the foregoing. I differ from van Fraassen (2002,

115–6), who at least suggests that "inertial mass" was somehow tacitly ambiguous before 1905.

22. See my discussions elsewhere (2004, 2009a, 2009b, 2000, and 2017).

REFERENCES

Braun, David, and Theodor Sider. 2007. "Vague, So Untrue." *Nous* 41: 133–56.

Bromberger, Sylvain. 2012. "Vagueness, Ambiguity, and the 'Sound' of Meaning." In *Analysis and Interpretation in the Exact Sciences. Essays in Honor of William Demopoulos,* edited by M. Frappier, D. Brown, and R. Disalle, 75–94. Dordrecht, the Netherlands: Springer.

Bureau International des Poids et Mesures (BIPM). 2012a. *Guide to the Expression of Uncertainty in Measurement.* 3rd ed. Sèvres, France: Joint Committee for Guides in Metrology. http://www.bipm.org/en/publications/guides /gum.html

Bureau International des Poids et Mesures (BIPM). 2012b. *International Vocabulary of Metrology,* 3rd ed. Sèvres, France: Joint Committee for Guides in Metrology. http://www.bipm.org/en/publications/guides/vim.html

Tal, Eran. 2011. "How Accurate Is the Standard Second?" *Philosophy of Science* 78: 1082–96.

Tal, Eran. 2012. *The Epistemology of Measurement: A Model-Based Account.* PhD diss., Graduate Department of Philosophy, University of Toronto.

Teller, Paul. 2000. "Twilight of the Perfect Model Model." *Erkenntnis* 55: 393–415.

Teller, Paul. 2004. "The Law Idealization." *Philosophy of Science* 71: 730–41.

Teller, Paul. 2009a. "Fictions, Fictionalization, and Truth in Science." In *Fictions in Science: Philosophical Essays on Modeling and Idealization,* edited by M. Suárez, 235–47. New York: Routledge.

Teller, Paul. 2009b. "Provisional Knowledge." In *Constituting Objectivity: Transcendental Perspectives on Modern Physics,* edited by Michel Bitbol, Pierre Kerszberg, and Jean Petitot, 503–14. Dordrecht, the Netherlands: Springer.

Teller, Paul. 2011. "Two Models of Truth." *Analysis* 71: 465–72.

Teller, Paul. 2013. "The Concept of Measurement-Precision." *Synthese* 190: 189–202.

Teller, Paul. 2017. "Modeling Truth." Philosophia 45: 143–61.

Van Fraassen, Bas. 2002. *The Empirical Stance.* New Haven, Conn.: Yale University Press.

12

LET'S TAKE THE METAPHYSICAL BULL
BY THE HORNS

BAS C. VAN FRAASSEN

TELLER'S CRITIQUE, which focuses on the identification of quantities, goes well beyond the problems that bedeviled the twentieth century "representational" theory of measurement. Equally, it stops nowhere short of the obvious "open question" response to the "analytic" theory of measurement: "Just what are those quantities that are postulated to be real?"

At the same time, lest the entire discussion should seem to fall squarely in a context of traditional metaphysics, Teller adds a disclaimer. Quantities, as well as relations and properties generally, are universals, but Teller moves his challenge outside the venerable problem of universals: "My claim is that the term, 'the temperature of the water in this glass,' does not have a referent. My reason is not in any way metaphysical" (chapter 11).

Teller adds that he is not taking sides on the metaphysical question of whether quantities in general, or the specific quantities dealt with in physics that he takes as examples, are or are not real. He writes, "I am not claiming that there are no quantities with exact values in nature," but he adds in a note "Nor would I claim that there *are* such things 'in nature,' whatever that might mean." The reality of quantities—or to put it in formal mode, the question of what (if anything) such quantity terms as "velocity" or "mass" have as referent—is not going to be assimilated to the problem of universals. Though philosophical, this question will be broached in issues actually faced in scientific practice as well as in the lay understanding of scientific modeling.

Empiricism and pragmatism meet here in a desire for the irrelevance of metaphysics. But it seems to me that we cannot very well wave away all questions about the reality of quantities when the focal question is about the referent of quantity terms or property terms. Such questions seem to arise very close to home. *Is there, was there ever, such a quantity as Galileo's force of the vacuum?* Or, nearer to home yet, when I explain Einstein's relativity revolution in class, it seems natural for me to say things like this:

> Some quantities that classical physics takes as fundamental, such as length, duration, and mass, lost their status. There really is no such thing as, for instance, the duration of a rocket's flight, as classically conceived. The real quantity is the space-time interval between departure and arrival. As it happens, a space-time interval between events during its flight (while unaccelerated) is measured by the rocket's own clock. But the classical quantity of time, equally measured by all clocks anywhere—there really is no such quantity at all.

"There is," Quine taught us, is the vernacular for "exists" and equally for the vernacular "is real." Thus, in my talk to the class, certain classically conceived quantities are said not to be real, unlike the quantity whose values are the magnitudes of space–time intervals. But what is this *being real?*

Philosophers aplenty have answers ready-to-wear. Teller and I agree in wanting most of those answers to be irrelevant. For example, some following David Lewis, more or less, would tell us that a *real* universal's extension is a natural class. Boundaries between natural classes are the lines along which nature is carved at the joints. The distinction between natural and unnatural classes saves metaphysical realism. The contemporary realist may consistently and piously point to "our best science" as discovering exactly which postulated or imagined quantities are real universals, with the claim that it is in our best science that reality will be carved at the joints.

Would any of this give us a clue as to what is meant by the assertion that Einstein's fundamental quantities rightly replace those of classical physics? However superficial or quick my in-class introduction to relativity may be, it is certain that a proper follow-up will expand on measurement, and not on Socrates's butcher's metaphor.

For what did Einstein point out about the classically conceived length, time, and mass? With the assumption, central to his view, that the speed of light is a universal constant, disruptive consequences followed for classical measurement operations for those quantities. Choose standard rods and clocks, and measure spatial or temporal extension at a distance: the relative motion of the measurement setups results in *discordant* measurement outcomes. Without *concordance* in different proper measurement results for the same quantity, *there is no such quantity*. Or, to avoid red flags waving before metaphysicians' eyes, say it this way: the classical theory fails to meet the requirement of empirical grounding.

Teller, following Tal, gives appropriate place to the requirement of concordance in the special garb of robustness, explained so as to take approximation into account: "The robustness condition is essential to the success of so proceeding. The condition functions as a prescription to check—with great thoroughness—that the various ways in which we assign values to quantities as described in our theories all fit together well enough not to engender difficulties" (chapter 11). To align our two ways of pointing to this requirement of concordance, read "ways in which we assign values" as "operations theoretically classified as measurements." The point made by Tal and Teller has its place in a larger historical context. I use the term "empirical grounding" (see, e.g., van Fraassen 2012, 2014). With this term I refer to the central requirement on the empirical sciences that Hermann Weyl spelled out in his classic *Philosophy of Mathematics and Natural Science:*

> 1. **Concordance.** The definite value which a quantity occurring in the theory assumes in a certain individual case will be determined from the empirical data on the basis of the theoretically posited connections. *Every such determination has to yield the same result* . . . Not infrequently a (relatively) direct observation of the quantity in question . . . is compared with a computation on the basis of other observations . . .

> 2. It must in principle always be possible to determine on the basis of observational data the definite value which a quantity occurring in the theory will have in a given individual case. (1927/1963, 121–22)

Teller would certainly have emphasized to Weyl that the "in principle" signals an extreme degree of idealization, and that to determine a value of some quantity through measurement can only be a matter of determining some bounds on the value within significant margins of error. Weyl would readily have agreed.[1] Let us take that as said, and return to it later. If we accept Weyl's dicta as specifying what is required for empirical adequacy, and take empirical adequacy as replacement for truth as the criterion to assess scientific success, then the metaphysical question about the reality of quantity terms falls away. For the only reference needed for a theoretical quantity term is a well-defined element of the algebra of quantities in each model of the theory. Assessment of the theory will not pertain to relations between such a quantity term and what there is in nature, but to the relations between the theory's models and measurement results. There is no metaphysical residue.[2]

But can Teller's critique perhaps be turned against this empiricist account? For while such an empiricist account removes the question of the reality of, for instance, the temperature of the water in this glass (or equivalently, the existence of a referent in nature of the term "the temperature of the water in this glass"), it hinges on the feasibility of humanly realizable measurement operations properly pertaining to theoretical models—I mean, to such models as may be offered, within the domain of a given theory (for, e.g., temperature in a glass of water, or the shape of that glass, or its mass).

There is, without question, the problem of achieving some degree of precision. Teller quotes VIM: " 'Even the most refined measurement cannot reduce the interval [that can reasonably be attributed to the measurand] to a single value because of the finite amount of detail in the definition of a measurand.'" Quite so. But concordance in measurement outcomes is something we can ask for at any level of precision. Suppose that two different methods are used to measure the mass flow rate of a gas. Whether the outcomes of the two procedures are concordant does not depend on how precise they are; it requires consistency, which can be as appropriate and clearly defined for intervals or fuzzy sets as for numbers.

Automobiles today offer a simple, practical example. To regulate the fuel for internal combustion requires, in today's electronically governed engine, data about the mass flow-rate of the air intake. One sort of mass flow sensor is thermal ("hot wire"); there are various mechanical sorts as well, such as the vane flow meter and the Karman vortex meter. Teller is right: the sensor signals cannot reasonably be thought to discriminate between one real number and nearby real numbers, or between any small real number interval and equally sized overlapping ones. But there is no difficulty in assessing whether the signals, obtained in these various ways, are consistent with each other. Notice, though, the question of principle in the background, behind the automobile design: What is the connection between the thermal sensing operation, which measures differences in temperature, and the mechanical one, which relies, for instance, on the rotation of a small vane? The connection is one *in a model of the fluid* or gas, whose equations relate the various theoretical quantities to one another.

What Teller denies in his critique is neither that quantity terms are precisely and definitely connected, in a theory, to specific elements of models, nor that such models are used by, for example, the automobile industry to represent certain everyday phenomena of practical concern, nor that measurement procedures are realized in physical form following a design dictated by this modeling. What Teller denies is the metaphysical "traditional realist" story that glosses this practice:

> The current proposal is not to scrap the concept of traditional accuracy realism in favor of some substantially different concept. Rather, I am urging a change in how we think about the concept. We apply the familiar concept but no longer in a traditional realist spirit. Instead we appreciate its status as an idealization. (chapter 11)

But can we really just stop with a comfortable acquiescence in the traditional metaphysics by ruling it a practically useful fiction or idealization? I am left dissatisfied if left without an alternative account that takes the problems seriously, even if I will be equally dismissive of the traditional answers. What can satisfy alone is an alternative account that answers the questions about identification and reference directly, without buying into realist metaphysics but as directly as the realist.

What the empiricist account adds is a rationale for declaring the metaphysical gloss to be irrelevant, and to show, without entering upon questions of "reference in nature," the form that assessment of success in science takes, properly understood, so as to bypass metaphysical questions.

In summary then, I cannot rest with Teller's practical and pragmatic acquiescence in the use of scientific language without a way to make sense of truth and reference for that language. The alternative to Teller's reconciliation of our use of theoretical quantity terms with his far-reaching critique is an account that provides direct answers to questions of reference and truth. It is the account that treats the language of science as *semi-interpreted*. Quantity terms do have referents: they refer to specific elements of theoretical models. To say that these terms are theory-laden, or that their reference is context-dependent, means just this: they refer to elements of models, and their connection with concrete natural phenomena is indirect.

The "cash value" of that connection lies entirely in the relation of those models to their targets, when offered as candidates for representation of those phenomena. That relation to the phenomena in turn is spelled out entirely in the theoretical classification of certain operations as determining values or ranges of value for those quantities. It is the theory itself, or the theory in the context of a larger background in science, that alone can tell us what the relevant measurements are. That those operations are measurements at all is something we see only through theory-colored glasses. After that, judgments of robustness or concordance are themselves true or false, and are checked by comparison of observable outcomes. In this way, scientific theories are understood as genuinely empirical and about the natural world, without their quantity terms understood as referring to something in nature, without metaphysical residue.

NOTES

1. See Weyl's section 79, "Measurement." In the same section of this book, first published in 1927, Weyl also emphasizes his awareness of the theory-dependence of measurement: "any such quantitative determination . . . is possible only on the basis of *theories*" (142).

2. In a more general form this point is of long standing in the semantic approach to science, as well as in, for example, Ronald Giere's related "model-based view." Thus, Giere said, "On a representational conception of models, language connects not directly with the world, but rather with a model, whose characteristics may be precisely defined. The connection with the world is then by way of similarity between a model and designated parts of the world" (1999, 56).

REFERENCES

Giere, Ronald N. 1999. "Using Models to Represent Reality." In *Model-Based Reasoning in Scientific Discovery*, edited by L. Magnani, N. J. Nersessian, and P. Thagard, 41–57. New York: Kluwer/Plenum.

van Fraassen, Bas C. 2012. "Modeling and Measurement: The Criterion of Empirical Grounding." *Philosophy of Science* 79: 773–84.

van Fraassen, Bas C. 2014. "The Criterion of Empirical Grounding in the Sciences." In *Bas van Fraassen's Approach to Representation and Models in Science*, edited by W. J. Gonzalez, 79–100. Dordrecht, the Netherlands: Springer Verlag.

Weyl, Hermann. (1927) 1963. *Philosophy of Mathematics and Natural Science*. Translated by Olaf Helmer. New York: Atheneum. First published in German in 1927.

13

TAKING THE METAPHYSICAL BULL BY THE HORNS
Completing the Job

PAUL TELLER

THIS NOTE IS A FOLLOW-ON to van Fraassen's "Let's Take the Metaphysical Bull by the Horns," commenting on my contribution to this volume. I shall give my own interpretation of what I think van Fraassen has in mind. I will then explain why I differ from him—and this, in two respects. I will explain why I am inclined not to follow the additional step that van Fraassen wants to take, and then I will consider van Fraassen's at least apparent characterization of my pragmatism as a kind of fictionalism. I do not think that it has to be taken that way. Explaining why will give me the opportunity to expand just a little on the last section of my paper.

INTERPRETING VAN FRAASSEN'S COMMENTS

A clear picture requires a brief synopsis of van Fraassen's constructive empiricism.[1] Van Fraassen draws a distinction between belief and acceptance as attitudes toward theories. "Acceptance of a theory involves as belief only that it is empirically adequate" (van Fraassen 1980, 12). But acceptance involves more than such belief:

> To accept a theory is to make a commitment, a commitment to the further confrontation of new phenomena within the framework of that theory, a commitment to a research programme, and a wager that all relevant [observable] phenomena can be accounted for without giving up that theory. (van Fraassen 1980, 88)

Belief, by way of contrast, is an unqualified epistemic attitude. To believe a theory is to take it to be true without qualification:

> For belief . . . all but the desire for truth must be "ulterior motives." *Since therefore there are reasons for acceptances* [e.g., the commitments mentioned previously] *which are not reasons for belief, I conclude that acceptance is not belief.* (van Fraassen 1989, 192, italics added)

Next we need van Fraassen's notion of what counts as observable: that of which oneself or anyone in one's epistemic community can become perceptually aware without the use of instruments. The fact that the test refers to our whole epistemic community avoids the usual problems with the observational/nonobservational distinction. The relativization to one's epistemic community is well motivated by the rationale for the whole position, as explained in my work (Teller 2001, 125–26, 129–31).

Van Fraassen provides a statement of constructive empiricism in *The Scientific Image:*

> Science aims to give us theories which are empirically adequate, and acceptance of a theory involves belief only that it is empirically adequate. (1980, 12)[2]

If beliefs that arise from an accepted theory are to be restricted to unaided observation, what role will instruments have? Rather than providing "windows" on a world that we cannot directly observe, van Fraassen regards instruments as producing new observable phenomena. Thus, a spectrometer produces observable spectrographs. A voltmeter produces observable phenomena such as its pointer pointing to "5" on the output display (cf. Teller 2001, 130).

Van Fraassen takes from Duhem the idea of thinking of theories as producing useful ways of classifying observable phenomena. Thus, the observable phenomenon of the pointer on the voltmeter pointing to "5" may get classified as "current with voltage of 5 volts." A visible streak on a photograph of a cloud chamber may get classified as "passage of an alpha particle." No belief in voltages or alpha particles is involved. Instead, to accept a theory is, among other things, to accept the usefulness of the theoretical classifications in accounting for the observable.[3]

VAN FRAASSEN'S REINTERPRETATION OF MY PRAGMATIST
CRITIQUE OF TRADITIONAL MEASUREMENT ACCURACY REALISM

I can now quickly apply this sketch of constructive empiricism to how van Fraassen seems to understand my pragmatist critique, what he wants to add, and why. Consider a voltmeter hooked up to a circuit. Observable characteristics include the pointer pointing to "5" and a big greasy fingerprint smudge. At the crudest level our theories will classify the pointer reading as important relative to certain interests and the smudge as uninteresting (or will not classify it at all). More specifically, and usefully, theories may classify the pointer reading as in the theoretical category of a voltage reading, with the value of five volts. Various of our theories may yield further classifications, such as the five-volt classifications being relevant to such and such, as leading to expectations of this and that, as recommending certain manipulations to achieve various ends, and so forth.

It is to be emphasized that constructive empiricists will regard these classifications as just that—with no reference to a quantity, a voltage, a value of five volts, or the like. Constructive empiricists may treat all of this with much the same affect of involvement one expects from realists, thinking in terms of voltages and so on but not as things believed, but as useful ways of thinking about and classifying observable things in application of relevant theories (cf. Teller 2001, 127).

My gloss so far may leave one feeling that van Fraassen's appeal to theoretical classification and its uses sounds somewhat like what he has characterized me as offering—"a practically useful fiction or idealization." What is distinctive about van Fraassen's view is the felt need for something more. Acceptance of the relevant theories and their classificatory categories must be filled out with *beliefs* in the empirical implications that we draw from our application of the theories and their classifications. These beliefs provide "[a] way to make sense of truth and reference" in ways that show "the form that assessment of success in science takes." The observable end products of the application of theoretical classifications provide the kind of referents and truth-evaluable judgments that constructive empiricists are ready to embrace as full-fledged beliefs.

WHY I DO NOT TAKE THIS ADDITIONAL EMPIRICIST STEP

To explain my attitude I will need to use a notion that has things in common with van Fraassen's acceptance but also differs in important respects. So I need a different term: *adoption*. To adopt a statement is to treat it as true for certain purposes. These purposes may be very specific and narrow. For instance, we can treat water as an incompressible, continuous medium for describing the fluid properties of water. The limiting case would be to assume a statement for the purposes of a reductio argument. Or the purposes for which one adopts a statement may be extremely broad, may be part of a background set of adopted statements presumed to apply unproblematically for anything that might reasonably come up: New York City is located in the United States.

Like adoption, acceptance has a pragmatic side; however, as I understand it, acceptance is limited to treating a theory as a basis for investigating a certain range of observable empirical phenomena. Adoption is pragmatic through and through: an instrumental attitude toward any statements that are entertained. Unlike acceptance, adoption does not require belief in empirical adequacy. And unlike acceptance, adoption of a statement presupposes that the statement can be refined, either by being made more precise (in the sense of being less vague), or more accurate, or both. (The reason for this additional clause is to facilitate integration with my larger view on treating truth, that I will mention later.[4])

In other words, I do not add van Fraassen's additional empiricist gloss to my treatment of accuracy because I adopt much but—in van Fraassen's sense of belief in unqualified truth—believe nothing.[5]

I can explain why most clearly where the disagreement is most salient, at the level of perception. Unlike van Fraassen, I take perception to be so much more modeling, not (when it is successful) a veridical grasp of a perceived world. This is a fundamental difference between myself and van Fraassen, who rejects representational theories of perception and indeed rejects mental representation of any kind (van Fraassen 2008, 24). I have written about perception as modeling in many places (in most detail in Teller 2008, section 7). But briefly, color vision provides an exemplar.[6] Naïvely, to see something as red is to become perceptually aware of the thing as having the intrinsic color property *red*. At least since the early modern period, people have appreciated that there are no intrinsic color properties, that they are "secondary properties" that involve the perceiver and, we now appreciate, also the immediate environment. In particular, a color cannot be understood as the surface spectral reflectance of an object—that is, the pattern of just what wavelengths of light an object will reflect.[7] Furthermore, our current theoretical understanding of color perception is itself incomplete, is not completely accurate, and is idealized in many ways.

The upshot is that any understanding we have of color, from the simple naïve idea of intrinsic properties to the best theory we now have, is not the identification of some property or relation in nature but a simplified model in terms of which we can usefully manage our color experience. I submit that it is going to be this way for all of perception. The world is too complicated for the brain to be able to afford to get it exactly right. Rather, the brain cuts corners and employs heuristic simplifications. In short, our perceptual experience of the world represents the world that we perceive in terms of simplified models. This includes our perceptual experience of discrete perceived objects. The problem of indeterminate spatial and temporal boundaries of objects is best understood as an idealized simplification that our perceptual apparatus makes for us that provides an excellent compromise between effectiveness and falling short of complete accuracy in representing a world that is unmanageably complex.

It is not such a big leap from perception to our representation of any aspect of the world, whether through perception, science, or other indirect means. All of it involves shortcuts, simplifications, and useful distortions. Any statement that we entertain is, at best, a representation that stands to be corrected and/or refined. Thus, the attitude of adoption is much more deeply qualified than taking one's judgments, generally, to be fallible.[8]

I should note that the forgoing is really a simplified presentation of my views. I also interpret the vagueness of language to be the flip side of our inability to get things both perfectly precisely and perfectly accurately. One can always eliminate inaccuracy of a very precise statement by making the statement less precise. One can get a little more sense of this from the final section of my paper. In (2017) I work this out in a lot more detail, including a way of thinking about truth that makes all of this work smoothly together.

FICTIONALISM?

At one point van Fraassen suggests that my attitude toward all the contents of what I adopt counts as "practically useful fiction or idealization." Idealization, yes, but I take exception that this counts as a kind of embracing of fictions. That a representation gets some things wrong does not automatically make it a fiction. Consider the opening line from the old time radio show *Dragnet:* "Ladies and Gentlemen. The story you are about to hear is true. Only the names have been changed to protect the innocent."[9] Medical and especially psychiatric reports will falsify irrelevant personal details to ensure the anonymity of the subject. But that does not make such a report a work of fiction.

Engaging with the world through telling but imperfect idealizations still constitutes knowledge of the world. For a first pass at understanding how this works, consider your efforts to become familiar with the physical features of someone's face by using photographs taken from various angles, none of which are in perfect focus. Still, you can use the photographs to build up an increasingly informative composite understanding of the face's physical features. One easily accepts this way of developing understanding through imperfect representations because one supposes that there is the objective face "out there" with which one could, at least in principle, make comparisons. One supposes that the objective face sets the respects in which the imperfect representations are and are not accurate.

I think that the foregoing is a good summary of how we proceed, but not a good gloss on how to understand what we are doing. The perceptions with which one checks the accuracy of the imperfect photographs are themselves only more representations. And we *know* that any perceptions to which one might appeal as setting the referent are themselves far from completely determinate and far from perfectly accurate. We can make sense of this without appealing to an objective face and without supposing an ultimately perfect representation. We do this by appreciating that each and every one of our representations is subject to refinement, in the sense that it could be superseded by another representation that better performs for us in all the ways than did the first, and from the point of view of which the prior representation was less precise and/or less accurate.[10]

When considering the face in question, there is nothing of which we can know exactly what it is. This might be because, when it comes to the face, there is nothing "out there in nature" that is in question. Or it might be because there is something, but something of which, because it is too complicated, we can have access to no exact representations—that is, representations that do not admit, at least in principle, the sort of refinement described in the last paragraph. This is a contingent circumstance, due to the fact that the world is just too complicated for us limited beings to produce representations that could not be refined.

The best we can do is to build up an ever-sharpened understanding by considering a collection of ever increasingly refined (in the sense sketched two paragraphs back) and increasingly integrated collection of imperfect representations.[11] Given that we are limited, contingently, to this kind of way of knowing about the world, the issue of the targets of the representations "out there in nature" becomes moot. (Dare I say, it becomes a matter of metaphysics?)

Given the complexity of the world, which limits our representational powers, this is the only possible kind of knowledge of the world, of the way the world is, really. Many have responded to this view by thinking of it as some kind of instrumentalism, or perhaps idealism. As with the epithet of fictionalism, I take such characterizations to be at best misleading. Our perceptual knowledge of the perceptible world is the exemplar, the gold standard of what it is to know about the world—we can say, redundantly, about what the world is *really*. When it turns out that all such perceptual knowledge is always limited and imperfect, always refinable, that is no reason to conclude that, after all, it is not knowledge of the world. The further knowledge that goes beyond perception is of the same kind in its limitations and in its refinability. There is no reason that it should not likewise count as knowledge of the world, although imperfect knowledge (we can say, redundantly, of the way the world is *really*). So this view should count as realism, in particular about both perception and science. Indeed, given the limitations on human knowledge, this is the only sensible way of understanding realism about any kind of human factual knowledge of the world.

NOTES

1. For the most part my references to van Fraassen's work will be indirect by citing my own work (2001), where I develop a much more thorough exposition of constructive empiricism. Copious references to van Fraassen's work will be found therein.

2. Note that constructive empiricism is a way of regarding science, a way of understanding what science aims to do. It does not rule out someone who embraces constructive empiricism from embracing additional beliefs about the unobservable but regards them as "supererogatory" (van Fraassen 1994, 182).

3. This idea is not discussed in my work (2001), nor have I found very extended discussion in van Fraassen's writing, but it pretty clearly plays an important role in his thinking. See van Fraassen (1980, 58; 2008, 139, 143, 164, 203, 261).

4. Van Fraassen's notion of acceptance was the sensible one for constructive empiricism in 1980, given the monolithic attitude toward theories that we all had at that time. But today we appreciate that theories all yield only local models that have limited application and that are never both completely precise and completely accurate. Given this shift in how we understand theories, van Fraassen's notion of acceptance may be under some pressure to embrace the broader pragmatic attitude of adoption (while retaining the clause that mandates belief in the empirical implications of a theory).

5. With the exception of sentential logic, combinatorics, and finite mathematics generally.

6. Giere (2006, chapter 2) presents a superb and accessible exposition of the relevant (known!) facts about color vision.

7. Giere (2006, 25–27) provides the argument why not.

8. Again, finite mathematics provides a plausible exception

9. Go to "Red Light Bandit," *Dragnet*, NBC Radio, July 14, 1949 (http://www.youtube.com/watch?v=R-Hp7D4SQ7A) for an audio!

10. The idea of a perfect, "limiting" representation as a regulative ideal makes perfectly good sense.

11. As I mentioned in the final section of my essay in this volume, Tal's robustness condition provides an example in a limited domain of the kind of thing that is in question here.

REFERENCES

Giere, Ronald N. 2006. *Scientific Perspectivism.* Chicago: University of Chicago Press.

Teller, Paul. 2001. "Whither Constructive Empiricism?" *Philosophical Studies* 106: 123–50.

Teller, Paul. 2008. "Of Course Idealizations Are Incommensurable!" In *Rethinking Scientific Change and Theory Comparison: Stabilities, Ruptures, Incommensurabilities?* edited by Léna Soler, Howard Sankey, and Paul Hoyningen-Huene, 247–64. Dordrecht, the Netherlands: Kluwer.

Teller, Paul. 2017. "Modeling Truth." *Philosophia* 45: 143–61.

Van Fraassen, Bas C. 1980. *The Scientific Image.* Oxford: Oxford University Press.

Van Fraassen, Bas C. 1989. *Laws and Symmetry.* Oxford: Oxford University Press.

Van Fraassen, Bas C. 1994. "Gideon Rosen on Constructive Empiricism." *Philosophical Studies* 74: 179–92.

Van Fraassen, Bas C. 2008. *Scientific Representation: Paradoxes of Perspective.* Oxford: Oxford University Press.

Nancy D. Cartwright is a professor of philosophy at the University of California at San Diego and Durham University in the United Kingdom. She is a philosopher of natural and social science who has worked primarily on issues of realism, modeling, causation, and evidence. She is a recent winner of the Lebowitz Prize for Philosophical Achievement, recipient of a MacArthur Fellowship, and is the author of *How the Laws of Physics Lie; Nature's Capacities and Their Measurement; The Dappled World: A Study of the Boundaries of Science; Hunting Causes and Using Them: Approaches in Philosophy and Economics;* and coauthor, with Jeremy Hardie, of *Evidence-Based Policy: A Practical Guide to Doing It Better.*

Anthony Chemero is a professor of philosophy and psychology at the University of Cincinnati. His research is both philosophical and empirical; typically, it tries to be both at the same time. His work focuses on questions related to nonlinear dynamical modeling, ecological psychology, complex systems, phenomenology, and artificial life. He is the author of *Radical Embodied Cognitive Science* and coauthor, with Stephan Käufer, of *Phenomenology: An Introduction.* He is editing the second edition of *The MIT Encyclopedia of the Cognitive Sciences.*

Ronald N. Giere is professor of philosophy emeritus at the University of Minnesota. He is the author of *Explaining Science: A Cognitive Approach; Science without Laws; Scientific Perspectivism;* and coauthor, with John Bickle and Robert F. Mauldin, of *Understanding Scientific Reasoning.*

Jenann Ismael is a professor of philosophy at Columbia University. She is the author of *Essays on Symmetry; The Situated Self;* and *How Physics Makes Us Free.*

Tarja Knuuttila is an associate professor of philosophy at the University of South Carolina. She has been active as editor in chief of *Science and Technology Studies,* as well as holding research fellowships in Finland and Switzerland.

Andrea Loettgers is a philosopher of science at the University of Bern, Switzerland. In her research she addresses epistemological questions surrounding the modeling practice in interdisciplinary research contexts.

Deborah Mayo is a professor of philosophy at Virginia Polytechnic Institute and State University. Her books are *Error and the Growth of Experimental Knowledge,* a winner of the 1998 Lakatos Prize in philosophy of science; *Error and Inference: Recent Exchanges on Experimental Reasoning, Reliability, and the Objectivity and Rationality of Science,* coedited with Aris Spanos; and *Statistical Inference as Severe Testing: How to Get Beyond the Statistics Wars.*

Isabelle F. Peschard is an associate professor of philosophy at San Francisco State University. Her work in philosophy of science has focused on the practice of science, specifically on the ways in which the activities of modeling, experimenting, and simulating are intertwined, both helping and constraining each other, and on the role of values and judgments of relevance.

Joseph Rouse is Hedding Professor of Moral Science in the philosophy department at Wesleyan University. He is the author of *Knowledge and Power: Towards a Political Philosophy of Science; Engaging Science: How to Understand Its Practices Philosophically; How Scientific Practices Matter: Reclaiming Philosophical Naturalism;* and *Articulating the World: Conceptual Understanding and the Scientific Image.*

Paul Teller is professor emeritus at the University of California at Davis. He is the author of *An Interpretive Introduction to Quantum Field Theory.*

Bas C. van Fraassen is a professor of philosophy at San Francisco State University and McCosh Professor of Philosophy, Emeritus, at Princeton

University. He is the author of *The Scientific Image; Laws and Symmetry; Quantum Mechanics: An Empiricist View; The Empirical Stance;* and *Scientific Representation: Paradoxes of Perspective.*

Michael Weisberg is a professor and the chair of philosophy at the University of Pennsylvania, where he codirects the Penn Laboratory for Understanding Science and is Associate Director for Outreach of MindCORE. He is the author of *Simulation and Similarity: Using Models to Understand the World.*

Eric Winsberg is a professor of philosophy at the University of South Florida. He is the author of *Science in the Age of Computer Simulation* and the forthcoming *Philosophy and Climate Science.*

INDEX OF SUBJECTS